International Calibration Study
of Traffic Conflict Techniques

NATO ASI Series

Advanced Science Institutes Series

A series presenting the results of activities sponsored by the NATO Science Committee, which aims at the dissemination of advanced scientific and technological knowledge, with a view to strengthening links between scientific communities.

The Series is published by an international board of publishers in conjunction with the NATO Scientific Affairs Division

A Life Sciences	Plenum Publishing Corporation
B Physics	London and New York
C Mathematical and Physical Sciences	D. Reidel Publishing Company Dordrecht, Boston and Lancaster
D Behavioural and Social Sciences **E Applied Sciences**	Martinus Nijhoff Publishers Boston, The Hague, Dordrecht and Lancaster
F Computer and Systems Sciences **G Ecological Sciences**	Springer-Verlag Berlin Heidelberg New York Tokyo

Series F: Computer and Systems Sciences Vol. 5

International Calibration Study
of Traffic Conflict Techniques

Edited by

Erik Asmussen

Institute for Road Safety Research SWOV, The Netherlands

Springer-Verlag Berlin Heidelberg New York Tokyo 1984
Published in cooperation with NATO Scientific Affairs Division

Proceedings of the NATO Advanced Research Workshop on International Calibration
Study of Traffic Conflict Techniques held at Copenhagen, May 25–27, 1983

ICTCT Organizing Committee
E. Asmussen, U. Engel, C. Hyden, E. Imre, J. Kraay, N. Muhlrad, S. Oppe

52032504

ISBN 3-540-12716-X Springer-Verlag Berlin Heidelberg New York Tokyo
ISBN 0-387-12716-X Springer-Verlag New York Heidelberg Berlin Tokyo

Library of Congress Cataloging in Publication Data. NATO Advanced Research Workshop on International
Calibration Study of Traffic Conflict Techniques (1983 : Copenhagen, Denmark) International calibration study of
traffic conflict techniques. (NATO ASI series. Series F, Computer and systems sciences ; no. 5) "Proceedings of the
NATO Advanced Research Workshop on International Calibration Study of Traffic Conflict Techniques held at
Copenhagen, May 25–27, 1983"—T.p. verso. "Published in cooperation with NATO Scientific Affairs Division"—T.p.
verso. 1. Traffic conflicts—Research—Congresses. 2. Traffic accidents—Research—Congresses. 3. Traffic safety—
Research—Congresses. I. Asmussen, E. (Erik), 1924. II. North Atlantic Treaty Organization. Scientific Affairs Division. III.
Title. IV. Series. HE5614.N392 1983 363.1'2563 83-20336
ISBN 0-387-12716-X (U.S.)

Printing: Beltz Offsetdruck, Hemsbach; Bookbinding: J. Schäffer OHG, Grünstadt
2145/3140-543210

CONTENT

The concept of traffic conflict was initiated in the United States in the 60s and raised a lot of interest in many countries : it was an opening towards the development of a new tool for safety evaluation and the diagnosis of local safety problems. The need for such a tool was great, because of the many situations where accident data was either scarce, unsatisfactory or unavailable.

Development of Traffic Conflict Techniques (TCT) started simultaneously in the 70s in several European countries and new studies were also undertaken in the United States, Canada and Israël. The need for international cooperation was rapidly felt, in order to exchange data, compare definitions and check progresses. An Association for International Cooperation on Traffic Conflict Techniques (ICTCT) was therefore created, grouping researchers and safety administrators, with the aim of promoting and organising exchange of information and common practical work.

Three Traffic Conflict Techniques Workshops were organised, in Oslo (1977), Paris (1979) and Leidschendam (1982). A small scale international experiment of calibration of TCTs was also carried out in Rouen, France, in 1979, and five teams took part in it from France, Germany, Sweden, the United Kingdom and the United States; results of this first experiment were used as a basis for the present enterprise.

To be acknowledged as a safety measuring tool, traffic conflict techniques had to be validated in relation to traditional safety indicators such as injury-accidents. Validation turned out to be a difficult task, requiring a lot of effort on data collection and the design of adequate statistical methods, and results obtained were not fully decisive. The pooling of efforts in this field appeared necessary for further progress and a first step towards it was the calibration of all existing TCTs, i.e. a detailed comparison of definitions, procedures, and type of data collected, in order to be able in the near future to use extended data bases and draw a conceptual framework for validation.

The second International Calibration Study was therefore organised in Malmö, Sweden, in June 1983, during which ten different teams would be experimenting their own TCT, simultaneously and on the same locations. Video-films were taken as a reference for all data collected, and a data treatment procedure, including both statistical and detailed comparisons, was designed. A preliminary meeting took place in Copenhaguen on 25 - 27 May 1983, giving all participating teams and interested observers an opportunity to discuss in details existing TCTs, and to finalize and adopt final procedures for the Malmö experiment.

This book contains all the papers presented in Copenhaguen, as well as a summary of discussions and conclusions.

JOINT INTERNATIONAL STUDY FOR THE CALIBRATION OF TRAFFIC CONFLICTS TECHNIQUES

Introduction speech ICTCT Meeting Copenhagen, 25-27 May 1983 and Malmö, 30 May-10 June 1983

Prof. Erik Asmussen
Director Institute for Road Safety Research SWOV
Leidschendam, The Netherlands

Ladies and Gentlemen,

Traffic unsafety can be regarded as the whole of existing and potential critical combinations of circumstances, incidents (conflicts) and accidents in traffic and the individual and social consequences (damages) caused by them. The main feature of incidents and accidents is that they are always preceded by a critical combination of circumstances in traffic. Such critical combination of circumstances for example in a situation can be described as a situation wherein, with unchanged traffic behaviour and/or unchanged traffic situation, the interaction between man, vehicle, road traffic and environment leads to accidents (see Figure 1).

Without taking into account the emotional content of the word, we could simply speak here of a coincidence of circumstances.
Such a combination or coincidence of circumstances in a traffic situation is always preceded by decisions, which are jointly determining whether the combination of circumstances becomes critical or not. Such decisions may refer to the purpose and scheme of travel, the mode of transport, the speed of the car and the alertness of the road user (provoked traffic behaviour).

If in the situation of a critical combination of circumstances, anticipating or "normal" change of behaviour is possible, because the road user recognises the critical (combination of) circumstances in time, there is no problem at all.
If there is no anticipating behaviour, or this is not sufficient, an emergency manoeuvre is needed, for instance emergency braking or evasive action.
If the emergency manoeuvre is succesful, an incident or conflict is the result. If the emergency manoeuvre fails an accident or collision arises.

NATO ASI Series, Vol. F5
International Calibration Study of Traffic Conflict Techniques
Edited by E. Asmussen
© Springer-Verlag Berlin Heidelberg 1984

Figure 1. Model of the accident process

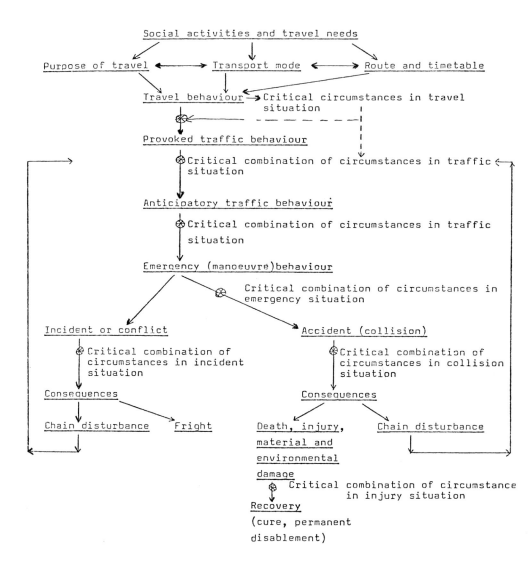

Both in the "anticipating" phase and in the "emergency" phase critical combinations of circumstances can affect the outcome.

I show you this phase model of the accident process, because we have to make clear to each other about which part of the process we are speaking if we use the word "conflict".

Before I shall speak about the importance and usefulness of traffic conflicts techniques, I want to make a comparison between the control of (the unsafety of) the transportation system and the steering of a fully loaded mammoth tanker.

If the wheel of such a vessel is swung right round, the effect (the output) will not become noticeable for some time. The slow response by the tanker is comparable with the slowness of accident registration.

The limitation of human perception abilities in noting slow (slight) changes is comparable with the limitations of statistical analysis methods for disclosing changes in the pattern of accidents.

The moment the changes in output are observed, it is often too late both on the tanker and in the transportation system to make effective corrective action.

Masters of giant tankers therefore do not respond so much to changes in the vessel's course (output variable; cf. accident statistics), but predict changes in output by responding to data on input and intermediate processes (input and process indicators), such as position of helm, speed, direction and speed of currents, etc. This is possible because they have sufficient knowledge and comprehension of the relationship between control variables and process variables, and the influence this has on changes in output. They do not wait, therefore, until the moment the output (change in course) manifests itself; they certainly do not wait till an accident has happened.

In research concerning shipping traffic, as well as in aerial traffic, the so called incidents or conflicts or near misses play an important role.

Of course that is also because accidents seldom happen, and if they happen, result in tremendous damage. But the main reason is that incidents or conflicts tell us about the critical combinations of circumstances in this traffic. In shipping and aerial systems, they even use this knowledge for training purposes. In research it is the most important source of information.

The challenge of this experiment that we are all concerned with, regarding the calibration of traffic conflicts techniques (TCT's), is to make clear the importance and usefulness of these techniques for the improvement of traffic safety. If we do not succeed in this, then we will fail regardless the interesting technical results. In many countries we find examples of the application of the TCT. However, the applicability is restricted and often restricted to experimental use. In various countries, however, there is a need for operational use on a larger and more general scale.

Sweden is one example of this. Mr. Mattson will give us a description of the background of this need. His problem as he states it in his paper is not so much the detection of dangerous locations but the analysis of the safety problem. The accident data are too scarce for a detailed analysis and the information stored in the accident report often misses the relevant cues to reconstruct what exactly did happen.

The Swedish conflict technique will be used to collect more information about the safety problems at specific locations. In The Netherlands, but I think also in many other countries, we feel the same need for additional information in order to make an analysis of traffic safety, and also in our country we look for a technique that is systematic and easy to use. In the USA, as can be seen from the paper of Mr. Migletz and Mr. Glauz, one is a little bit dissatisfied about the applicability of the TCT for safety analysis purposes. They lowered their aims and made the technique applicable in order to detect "operational deficiencies" as they call it. A concept that is related to dis-comfort and feelings of unsafety that also can be regarded as negative aspects of transportation. But also in the United States there is a need for such an easy-to-use technique to solve safety problems.

This brings us to the very heart of the problem: How relevant is the analysis of traffic conflicts for the analysis of traffic safety?

So far I mentioned two kinds of usage of the conflict analysis technique: The detection of dangerous locations and the diagnosis of the safety problem. An administrator, however, who is in charge of the safety of a road network, is primarily interested in the application of the technique with regard to the solution of the safety problem he has detected and analysed. He wants to know how to control safety. If the diagnosis leads to a conclusion about what is wrong at a particular location, then this does not lead directly to a solution of this safety problem. Various safety measures can be taken in order to solve the safety problem. It is not necessary that the application of these measures leads to a definite solution of the problem at hand. Measures often have side-effects. They may influence the situation in more than one way. Road surface improvements may attract traffic, traffic signals may cause changes in routes, etc. The improvement of road safety is a dynamic process that asks for constant evaluation of results.

Conflict analysis technique as a technique for quick evaluation of safety measures seems to me a very efficient tool to improve traffic safety in a dynamic way. This evaluation of safety measures is also urgent, because the diagnosis will always be uncertain and result in a hypothesis rather than an irresistable fact. Together with the uncertainty about the effectiveness of safety measures, this seems to ask for a short term evaluation of the effects. Only behavorial studies and especially systematic observation as can be found in a well "articulated" conflict analysis technique seem to give us a way out of this problem.

In practice, accident studies can hardly be used for this purpose.

The only justification for the use of TCT for the purposes mentioned can be found in a well-established theory about traffic safety. How do traffic accidents take place? Under what circumstances do traffic situations escalate into such a way that correction is not possible any more and an accident results.

Most of the conflict teams that are present to-day and will join us in the experiment, work on the basis of a more or less specific theory about this escalation. Elements of this theory can be found in their definitions of a conflict. Many teams use time as a basis to define the severity of a conflict. The less time there is left to react to a critical combination of circumstances, the more dangerous the situation is. But time is not enough. Manoeuvering space is also needed. And if we are primarily concerned with injuries or fatalities then also the kind of road-usage is very important.

As stated before, there is a need for a general technique that can be easily applied in various situations. Many techniques are rather specific, dealing with car-car conflicts only or car-pedestrians conflicts, conflicts at intersections with dense traffic, etc. Especially in this experiment the confrontation of many different points of view of various experts can lead to a fruitful discussion about the characteristics of traffic situations that lead to danger. In itself a conflict need not to be dangerous. Almost all conflicts can be dealt with adequately. It is important to find out which conditions are responsible for the loss of control in the rare cases the result is not a conflict, but an accident. In this respect, the conflict analysis technique can be regarded as a part of a general theory about traffic safety. We will not be able to solve all traffic safety problems at once with a magic formula called conflict analysis but if we look at is as part of a general theory about traffic safety, then, may be, this kind of systematic observation may help us to get more insight in safety problems.

Calibration and the discussion of the results is the first step in the development of a technique that is soundly based on a well-established theory. Confirmation of the theory by means of validation studies is a necessary second step that I hope will not be ignored. But also for this second step the calibration of techniques is valuable. It will give us a basis for comparison and discussion of results.

REVIEW OF TRAFFIC CONFLICT TECHNIQUE APPLICATIONS IN ISRAEL

A.S. HAKKERT, Road Safety Centre
Transportation Research Institute
Technion - Haïfa, Israël, 32000

Introduction

The traffic conflict technique (TCT) is widely believed to be a useful tool for objectively measuring accident potential at intersections and other hazardous locations [Williams 1981]. It can be used as a rapid evaluation tool for traffic engineering improvements and for learning about the possible contributory factors to accidents [Perkins and Harris 1968].

The basic relationship which should be underlying in all TCT studies is that there is a proven and consistent relationship between accidents and conflicts. This relationship has not, however, been satisfactorily established. Early studies [Perkins and Harris 1968; Spicer 1973] seemed to provide encouraging results but were of limited scope, and have more recently been criticized [Glennon 1977; Cooper 1973]. Possibly, one of the confounding factors in comparing conflicts and accidents is the general comparison of the two without differentiating between types and severities. This is difficult to achieve at single locations which might have only few accidents. Another aggravating factor which has not been sufficiently studied is the variability of conflict frequencies under similar conditions or at similar sites [Hauer 1978].

Accident severity varies according to circumstances. Serious accidents are associated with low-volume, high speeds and pedestrians and night-time alcohol consumption. Slight injuries are associated with high-volume, low speed (rush hour) conditions. Since conflicts can be regarded as a very light form of traffic disturbance, it would be reasonable to assume that they be differently related to accidents of varying severities.

This paper presents a review of the conflict studies which have been conducted in Israel. It discusses some of the results achieved and presents some of the questions which remain unsolved.

NATO ASI Series, Vol. F5
International Calibration Study of Traffic Conflict Techniques
Edited by E. Asmussen
© Springer-Verlag Berlin Heidelberg 1984

Comparison Between Objective and Subjective Measures of Traffic Conflicts

In 1979, Balasha reported on a study concerned with the objective definition of traffic conflicts [Balasha et al., 1980].

In analogy to car-following models describing the motions of following vehicles in a traffic stream, an attempt was made to define the motions of two vehicles following each other through an intersection. Development of such a model could possibly lead to the detection of unusual events, assessment of the difficulty of various vehicle manoeuvres, and to the possible identification of those locations at an intersection where difficulties in manoeuvring are encountered. The reactions and manoeuvres of vehicles on the approach to an intersection were studied. Manoeuvres were recorded continuously, so that a microscopic model of the traffic flow could be defined in the following way:

reaction = stimulus x sensitivity

Most car-following models of this kind deal with single lane traffic on an undisturbed straight section of highway, and define the reaction as a change in tangential velocity. The present model, however, extended the definition of reaction in order to handle interactions of pairs of vehicles on the approach to and through an intersection. For such cases, two dimensions of motion must be considered, and all terms of the model-reaction, stimulus and sensitivity, were defined accordingly. Two groups of variables were defined, one dealing with motion along the axis of travel, and the other dealing with angular motion, i.e., changes in direction of travel. The resultant reaction a_e was defined as:

$$a_e = \sqrt{a_T^2 + a_R^2}$$

a_T = change in tangential velocity
a_R = change in radial velocity

Two urban unsignalized intersections were filmed, using a Bolex H16 16 mm. film camera at a rate of 24 fps. The intersection area and the approach were marked with an orthogonal grid of 1 x 1 m. stripes. The film was analyzed using a Hadland Vanguard film analyzer, and in order to translate the film perspective to real coordinates, a polynomial regression of the coordinates was undertaken. The two intersections had fairly similar traffic flows. Each had a major road of 11 - 12 metre width. One (termed A) was an X-type crossroad. It had had 24 injury accidents in four years, had unmarked lanes, no pedestrian crossings, and a limited field of vision. The second intersection (termed B) was of a T-type with only 9 injury accidents in four years, well marked, with zebra crossings and a clear field of vision. About 139 vehicles were filmed on intersection A and about

208 vehicles on intersection B, resulting in the analysis of some 60,000 film frames.

For each individual vehicle, the following values were calculated: tangential velocity and velocity change, vehicle angle, angular velocity and velocity change, resultant velocity change.

For each pair of vehicles, a flow equation was calculated by means of regression analysis. Because of the delay between stimulus and reaction, various reaction times were assumed, and for each vehicle pair that equation was chosen which produced the highest correlation coefficient. The general motion equation was of the form:

$$\bar{a} = \lambda_{oo} + \lambda_{o1} \, \Delta s/Hd + \lambda_{o2} \, \Delta\theta \, \Delta\omega$$

where:

$$\Delta s = s_{n+1} - s_n \quad ; \quad Hd = x_{n+1} - x_n$$

$$\Delta\theta = \theta_{n+1} - \theta_n \quad ; \quad \Delta\omega = \omega_{n+1} - \omega_n$$

$\bar{\lambda}_{oo}$ – is the average of λ_{ooi} for individual pairs of vehicles conducting a similar kind of manoeuvre.

An irregular event was defined where the resultant deceleration exceeded $\bar{\lambda}_{oo}$ by some "safety margin and was therefore defined as $(a_e)_L$

$$|(a_e)_L| > \bar{\lambda}_{oo} + 2\sigma$$

In order to bring out those events that continue over a period of time, a further definition was introduced based on the sum of decelerations. A certain critical sum value was determined for vehicles involved in irregularities which, when exceeded, was termed a 'near accident.'

$$-\sum_1^F a_e > (-\sum a_e)_L$$

where:

$- a_e > (- a_e)_L$ for each film frame

$(-\sum a_e)_L$ – sum critical value of decelerations

On the basis of these criteria, determined for each intersection, 'near accidents' were selected and located within each of the two intersections.

Table 1 presents the number and percentage of vehicles exceeding the preset critical value.

Table 1: Number and percentage of vehicles exceeding critical value

	Intersection A	Intersection B
Total vehicles observed	139	208
Exceeding critical value	103	74
Percentage exceeding	74.1	35.6

The percentage of vehicles involved is large, particularly at intersection A. It should be emphasized that all vehicles which produced a value exceeding the critical value (even in one frame only) are here included.

Table 2 presents results of the number and percentage of vehicles involved in near-accident situations as defined above.

Table 2: Number and percentage of vehicles involved in near-accident situations and their critical cumulative values of resultant deceleration

Intersection	No. of vehicles involved in 'near accidents'		Critical value of sum of decelerations $\sum (a_e)_L$ – m/sec.^{2}
	number	percentage	
A	22	15.8	65
B	16	7.7	25

These vehicles contributed 49 percent of the total sum of deceleration in irregular manoeuvres at intersection A and 75 percent of intersection B.

In 1981, Shinar conducted an independent investigation on the film containing the 16 'near-accidents' at intersection B [Shinar 1983]. Shinar evaluated the 16 sequences by having observers subjectively rate the conflicts on a scale of 0 - 100, where 0 is a no conflict situation and 100 is a collision. Different groups of observers based their rating on two different definitions : combined lateral and longitudinal deceleration (termed SLLD - and similar to Balasha's objective definition) and a subjective time to collision (termed STTC). In addition, the observers viewed the 16 conflicts three times (in different order each time) to see if experience without feedback (which simulated experience in real life observations of conflicts) resulted in any learning effect. The 'objective' Balasha scale which also included severity of conflict was termed OLLD and results were compared.

Correlational analysis of the data indicated that : (1) observers remained very consistent in their judgments, the correlations among trials ranging from 0.96 to 0.99 for both SLLD and STTC; (2) the relationship between the subjective ratings and the objective scores was marginally significant, ranging from 0.48 to 0.66 for the correlation between OLLD and SLLD, and 0.43 to 0.51 for the correlation between OLLD and STTC; and (3) the relationship between the two subjective definitions was strong and yielded correlation coefficients of 0.82 to 0.94.

One example calculating the rank correlation between objective (OLLD) and subjective (SLLD) ratings is given in Table 3.

Table 3: Objective and subjective ratings

Balasha deceleration	Balasha ranking	Subjective rating	Subjective ranking
40	13	6.7	16
100	6	38.3	10
76	7	10	15
33	15	18.3	12
32	16	11.7	14
223	1	52.1	7
48	9	70.4	1
172	3	67.1	2
196	2	52.2	6
44	11	14.9	13
46	10	29.1	11
142	5	52.6	5
53	8	64.1	3
36	14	49.5	8
156	4	54.3	4
43	12	47.1	9

The rank correlation for the two rankings is $\rho = 0.67$.

These results seem to indicate that practice alone at watching conflicts does not tend to affect the judgments or produce 'learning.' Furthermore, the subjective evaluations of a conflict may be based more on an intuitive concept of a 'near accident' rather than on the formal definition the observer is given; hence the high correlations between the groups that were given different definitions. Finally, the use of subjective judgements appears to yield quite different results than objective scoring, indicating that either the formal definition is inappropriate, or incomplete, or that human observers cannot reliably estimate a 'conflict' on the basis of a theoretically derived definition. Since Shinar's study was conducted on data from only one intersection, it seems impossible to determine which measure - the objective or the subjective - is the more appropriate.

One limitation of the present comparison lies in the fact that objective and subjective conflicts were not independently selected. The objective selection of the 16 most severe resultant decelerations was used as the basis of the subjective rating. An interesting addition to the present study would be to have observers view the whole film and independently select and rate conflict situations.

A Conflict Study Evaluation of Flashing Amber Traffic Signal Operation

Mahalel et al., (1982) evaluated the safety, energy and environmental aspects of switching traffic signals to flashing amber during off-peak hours. To assess the safety aspect, they made use of the traffic conflict technique. Four urban signalized intersections were selected for their study, two 4-way and two 3-way intersections. These intersections were observed in the two modes of operation. Basic counting samples were of 15 minute length, during which traffic volumes on each approach were counted and the conflicts were recorded. The definition of conflict used in their study was "an event in which one road user causes another road user to change his course of travel in time or space." Both deceleration and acceleration caused by another road user were included as conflicts and so were changes in direction (swerving). This definition is different from that adopted at the Oslo Conference in that there is no mention of a risk of collision. The authors felt that they should remove as much subjective evaluation as possible, and also felt that their definition would increase the sample size of observable conflicts. The basic form, used to record the conflicts, is attached at the end of this paper.

Table 4 presents a summary of the data collected at the four intersections.

Table 4: Summary data of traffic volumes and conflicts at four intersections

	Regular operation	Flashing operation
Total hours of operation	34.25	29.75
Average number of vehicles per hour	829	789
Total number of conflicts observed	636	1054
Average no. of conflicts per hour	18.6	35.4
Percentage of conflicts:		
Rear-end	53	15
Crossing and merging	24	75
Pedestrians	23	10

It can be seen from Table 4 that traffic volumes were fairly similar during the two types of operation. The number of conflicts was much higher in the flashing than in the regular mode. However, there was a trade-off in the types of conflict. Whereas in the regular signal mode, a majority of the conflicts were of the rear-end type (53%), in the flashing mode the large majority of conflicts involved a crossing or merging manoeuvre (75%). The percentage of conflicts involving a pedestrian decreased.

The authors explain that whereas during regular operation, the pedestrian shares his right-of-way with a right turning vehicle, during flashing operation he is free to select the most opportune moment. From a more detailed study of conflicts in various ranges of traffic volumes, the authors established that the number of conflicts in the flashing mode rose rapidly from about 17 per hour at 500 veh/hour to about 36 per hour at 1300 veh/hour. During regular operation, the number of conflicts decreased from about 21 at 500 veh/hour to about 16 at flows above 700 veh/hour and did not vary much with flow.

They adopted a conservative decision rule whereby they recommended that the total number of conflicts in the flashing mode should not exceed the total number in regular mode plus one standard deviation. Thus, they arrived at a limiting value of 600 vehicles per hour entering the intersection from all directions to convert from regular to flashing operation. This is not such a radical departure from the accepted practice in Israel and other countries, whereby signals are switched to flashing amber at night when hourly volumes decrease below a value of 400 for four consecutive hours.

In reviewing this study, the basic dilemma remains that on the basis of the conflict study there is a trade-off between rear-end conflicts which decrease, and crossing and merging conflicts which increase during flashing operation. The question to be asked is whether the number of accidents is also expected to behave in each mode accordingly. It is very plausible to expect that the number of right angle and turning movement accidents will be greater in the flashing mode than in the regular mode of signal operation. It is also reasonable to expect that rear-end accidents will decrease. It is difficult to speculate as to the size of change on the basis of the conflict study. One would have to develop accident-to-conflict ratios for the various types of conflicts and for the various modes of operation, and it is exactly this kind of ratio which is lacking and difficult to obtain.

This brings us back to the basic question whether conflict studies on their own are sufficient basis for engineering decision-making on traffic operational measures. It would seem that on the basis of the findings in the study reviewed, and in view of the fact that from energy and environmental considerations, flashing signal operations are highly beneficial, the idea is worth a full scale before-after experiment.

Conflict and Exposure Study of a Hazardous Intersection

Within the framework of the rehabilitation of a low income residential area, Balasha carried out a total safety evaluation of the neighbourhood, [Balasha et al., 1982]. The central intersection adjacent to the local cinema had 7 injury accidents out of a total of 30 in the area during the 5 years 1977-81. In order to shed some light on the possible engineering deficiencies at that 4-way inter-section, a conflict study was undertaken. The conflict definition adopted by Mahalel et al., (1982) was used. Two observers counted traffic and conflicts during the hours of 14.20 to 19.00 on a midweek day. A total of 245 conflicts were counted during this 4:40 hour period. Most of the conflicts (65%) occurred on the major road. 76 of the 245 conflicts were with pedestrians, and again 71% occurred on the major road which has priority over the two minor approaches. A comparison of the conflicts with traffic and pedestrian counts between each pair of directions (V,P) showed that the following relationship explained 60% of the variation in conflicts:

$$C = -1.3 + 0.035 \, V + 0.013 \, P \quad (r = 0.73)$$

C = number of conflicts, V = number of vehicles, P = number of pedestrians crossing.

On the basis of the conflict study and further engineering evaluations, it was decided to turn the 4-way intersection into a T-type junction and turn one approach-adjacent to the cinema into a pedestrian mall.

It is felt that this single operational conflict study is too limited in scope to draw any major conclusions from it.

Conclusions

A number of techniques have been applied to the subject of conflict studies in Israel. All of these applications were carried out within the framework of research at University Institutes. It can therefore be stated that the conflict technique is not yet operational in Israel. Although there is some general interest in the TCT, its lack of application in Israel is probably connected with the following reasons: (1) not many engineers are familiar with the details of the technique; (2) it has not been recommended as a standard evaluation technique in traffic engineering manuals, handbooks or textbooks;

(3) not many other countries have adopted it as an operational technique. The aspects of TCT discussed in this paper illustrate a number of interesting points which still remain unsolved.

REFERENCES

Balasha, D., Hakkert, A.S., Livneh, M. (1980). 'A Quantitative Definition of the Near-Accident Concept." TRRL Suppl. Report 557, Transport & Road Research Laboratory, Crowthorne, U.K.

Balasha, D., Katz, A., Yogev, Y., Elgrishi, A. (1982). 'Traffic Safety in Residential Renewal Areas.' Publication 82-15, Road Safety Centre, Technion, Israel (In Hebrew).

Cooper, P.J. (1973). 'Predicting Intersection Accidents.' Ministry of Transport, Ottawa, Canada.

Glennon, J.C., Glauz, W.D., Sharp, M.C., Thorson, B.A. (1977). 'Critique of the Traffic-Conflict Technique.' TRR 630, Transportation Research Board, Washington, U.S.A.

Hauer, E. (1982). 'Traffic Conflicts and Exposure.' AAP Vol. No. 5, Pergamon Press, U.K.

Mahalel, D., Peled, A., Livneh, M. (1982). 'Safety, Energetic and Environmental Evaluation of Flashing Signals at Off-Peak Hours.' Report 82-17, Transportation Research Institute, Technion, Israel (In Hebrew).

Perkins, S.R. and Harris, J.I. (1968). 'Traffic Conflict Characteristics - Accident Potential at Intersections.' HRB Rec. 225, Highway Research Board, Washington, U.S.A.

Shinar, D. (1983). 'Subjective vs. Objective Measures of Traffic Conflicts.' To be published.

Spicer, B.R. (1973). 'A Study of Traffic Conflicts at Six Intersections.' TRRL Rep. LR 551, Transport & Road Research Laboratory, Crowthorne, U.K.

Williams, M.J. (1981). 'Validity of the Traffic Conflict Technique.' AAP Vol. 13, pp. 133-145, Pergamon Press, U.K.

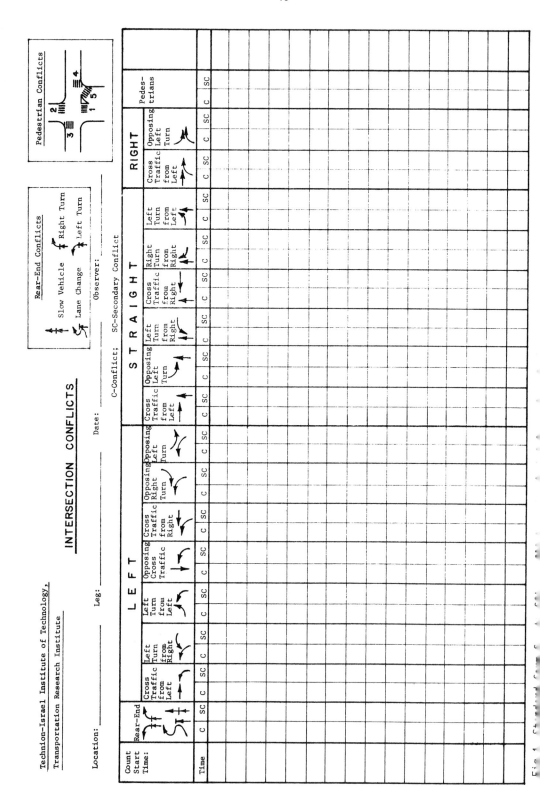

Fig. 1. Standard form for the collection of intersection conflict data

CONFLICT OBSERVATION IN THEORY AND IN PRACTICE

V.A. Güttinger
"Advisie", Consultancy firm for Government and Management
Nieuwe Uitleg 26, 2514 BR The Hague, The Netherlands

1. Introduction

When speaking about the dangers of traffic, most of us think of accidents. In the
Netherlands however there is a tendency towards a broader view of this safety

concept. The traffic safety in this broader opinion is not only limited to accidents
but also has to do with conflicts and feelings of fear.

Accidents describe only a part of the traffic safety problem. However, they still are
serious events that have to be avoided.

The changing opinion of the safety concept partly has to do with opinions about the
psychological importance of fear: people not only should be safe, they should feel
safe too.

For a part this changed conception has to do with the awareness that accidents are

rather unsatisfactory indicators for traffic unsafety:

- the registration of accidents is limited and not always reliable or complete;

- accidents, although they happen too often, are relatively rare events;
- the fact that accidents must take place before one can determine the risk of
  locations is, from an ethical point of view, a basic disadvantage.

These shortcomings of the accidents criterion lead to the search for a more frequently occurring and measurable phenomenon as a criterion for traffic safety.

2. Short history of conflict observation

The aforementioned problems with the accident criterion gave rise to the development
of so-called "conflicts techniques".

In its origin this development started after world war II. In aviation "pilot errors" or "critical incidents" were then used as measures of safety performance
(Fits & Jones, 1947; Flanagan, 1959).

The term "conflict" in traffic research was introduced by Perkins and Harris (1967).
An important refinement of the orginal technique of Perkins and Harris was introduced by Spicer (1971) with his concept of "severity grade". Most of the developed
techniques are based on his work. Some of the researchers in this area followed a

line introduced by Hayward with the "time-measured-to-collision" concept* (Hayward, 1972).

Despite the general agreement between the participants of the First International Traffic Workshop, about the definition of a "conflict"**, and on the main aspects of the operational definitions of conflicts (evasive or avoidance actions), there seems to be some confusion regarding the place of the conflict in the chain of events as illustrated in figure 1.

For some, the conflict is an event that precedes an evasive action that can be either succesful or not (collision). For others, it is the same as a near-miss situation after an evasive action. In this last view, a conflict cannot lead to a collision but is an event parallel with a collision.

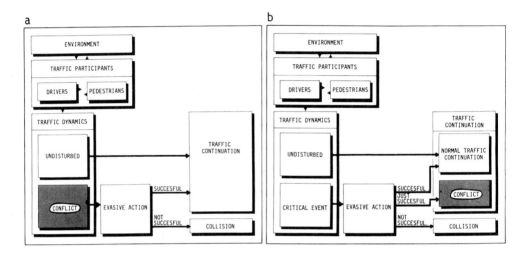

a) Conflict as potential accident.
   Evasive action - sometimes combined with distance between participants - indicates previous con-
   ·flict.

b) Conflict as near-miss situation.
   Just succesful evasive action (see
   text) - sometimes combined with
   distance between participants -
   indicates conflict.

Figure 1.   Place of the conflict in the traffic process.

---

*   Time-measured-to-collision (TMTC): "The time required for two vehicles to col-
     lide, if they continue at their present speeds and on the same path" (Hayward,
     1972, p. 9).

**   A traffic conflict is an observable situation in which two or more road users
     approach each other in space and time to such an extent that there is a risk
     of collision if their movements remain unchanged (Proceedings First Interna-
     tional Traffic Conflicts Workshop, 1979).

3. Conflict observation in the Netherlands

The development of our conflict observation technique was mainly inspired by the work
of Spicer (1971, 1972, 1973) and started in 1975, although it should be mentioned that
some conflict related research in the Netherlands has been done earlier (Frusch e.a.,
1971; Paymans, 1972).

More recently a technique is developed based on the TMTC-idea of Hayward (v.d. Horst &
Symonsma, 1980).

Our work aimed at developing a reliable and valid conflict observation technique that
could be used for the prediction of the safety of child pedestrians, consisted of four
steps.

We will not present the results of these phases (operationalisation; tests of reliability, applicability and validity) in detail here. They can be found in earlier contributions to the meetings of the ICTCT (Güttinger, 1977, 1980, 1982).

It turned out that our "serious conflicts"* were reliable observable phenomena (inter-
and intra-rater reliability) that showed a strong association with accidents of child
pedestrians (see appendix).

Given this fact and the fact that other variables (traffic volumes, subjective estimation of risks) in our research, were less succesful in predicting accidents, we
feel that the use of this technique is justified for those situations it is mentioned
for.

In the next paragraph some of the results of the application of the technique are presented with special attention for one way of application (the so-called "personal observation"). It gives some idea in which conflict observation can contribute in evaluating traffic situations and traffic designs of residential areas.

4. Conflict observation in practice

The developed technique is used in two ways:
    a) by means of sector observation and
    b) by means of personal observation.

ad a) The method of sector observation is especially suited for the determination of
       the risk of certain spots, e.g. an intersection or a part of the road. Most of
       the different developed techniques, or all of them, are used in this way.

---

* Serious conflict: a sudden motor reaction by a party or both of the parties involved in a traffic situation towards the other to avoid a collision, with a
  distance of about one metre or less between those involved.

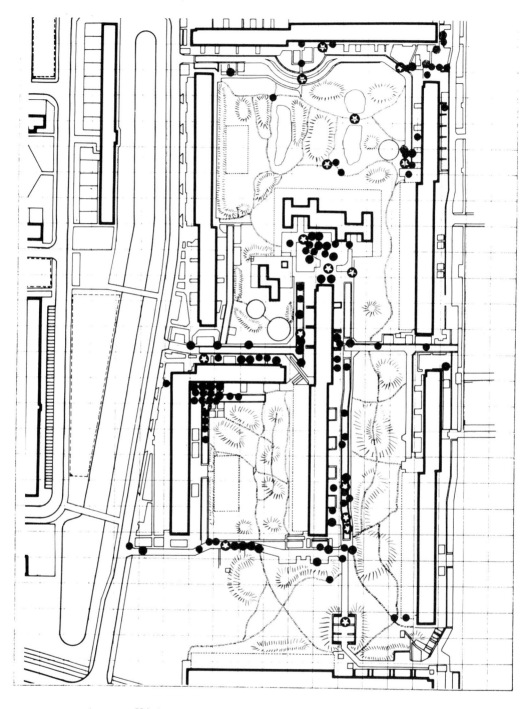

⊗ = serious conflict

Figure 2. Place of serious conflicts and encounters in a experimental area.

ad b) A special feature of our work is the application of conflict observation by
means of what is called "personal observation". In the case of personal observation, individual road users (pedestrians) are followed (by trained observers)
for a certain time or along certain routes. This method is suited for the comparison of larger environment units (e.g. neighbourhoods), for the detection of
high risk spots within large areas or to trace the relative risks of routes or
groups of pedestrians.

In the following some short examples of the kind of results achieved with the latter
method, will be given.

- *example 1*

In one of the first applications of the technique in two residential areas, an experimental area (one of the first residential yards) and a control area, children were
followed during their play after schoolhours.
Figure two shows the places where serious conflicts and other types of encounters were
observed in the experimental area.
Inspection of these places where serious conflicts occurred, showed that in most cases
the mutual visibility of pedestrians and traffic was hindered by measures that were
intended to enhance traffic safety (like obstacles).
Example two also shows the importance of the visibility and the incorrect estimation
of pedestrian behaviour by the designers.

- *example 2*

In recent research the application of the technique was extended to adult pedestrians.
In three areas, each with a different traffic layout, varying from minimum measures
to keep out traffic not belonging to the area, up to residential yards, children and
adult pedestrians were followed on their way through the areas.
The example concerns the residential yard. A feature of the residential yard is that
there is no strict separation between different traffic participants by means of a
separate road and sidewalks. Pedestrians are allowed to walk and play on the same
area that is used by traffic, although in most cases there is some form of walking
area.
In the present residential yard very few serious conflicts of children occurred,
spread over the area and when playing on the road.
Serious conflicts of adult pedestrians happened systematically at certain spots.
There is a kind of walking area at the side of the road. This walking area changes
from one side to the other, at points where chicanes are created (to reduce speeds)
by means of parking areas.
In contrast with children, who play and walk on the road (allowed), adult pedestrians
showed a tendency to traditional pedestrian behaviour: they used the walking area.
But at points where the mutual visibility is hindered by parked cars they are forced
to cross the road which results in serious conflicts (see figure 3).

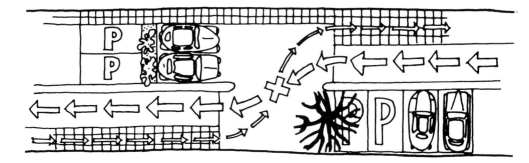

Figure 3. Pedestrian is forced to cross the road.

Remaining with this example an other advantage of the personal observation technique
can be illustrated.

As mentioned, very few serious conflicts with children occurred. Also in the other
areas, with less radical traffic measures, very few serious conflicts happened. In
fact, no differences between the three areas were observed.

However, for the residential yard children required more time to cover the same
length of path than the children in the other areas. They had less serious conflicts
(compared to children in the other areas) per unit of time, but (probably because of
the attractiveness of the area for play), they use more time for a comparable route
resulting in an equal amount of serious conflicts in all areas.

The last example is also an illustration of the advantage of the personal observation
method: the insight it gives to the exposure of traffic dangers.

- *example 3*

In the same research project we compared the follow-time of pedestrians that did and
did not have an encounter with traffic. The former required significantly ($p < .01$)
more time to cover the same route, than the latter.

It also turned out that among those pedestrians that had an encounter with traffic
the proportion of elderly people (> 60 years, a vulnerable group according to the
accident statistics) was much larger than could be expected when considering there
presence in the areas (pedestrian counts).

Conclusion: older people walk slower which enhances their exposure time which results in more encounters.

Perhaps the results of our conflict observations as presented in these examples are
not overwhelming. We know from accident data that mutual visibility is important in
preventing accidents. We know that older people are more vulnerable as pedestrians.
It does seem to show two additional things:

   1) a certain face-validity and concurrent validity of the developed technique
      (apart from the predictive validity);
   2) the use-value of the technique: faults in the traffic design can be readily
      detected.

LITERATURE

Fits, P.M. & R.E. Jones. Analyses of factors contributing to 460 "pilot-error" experiences in operating aircraft controls. Aero Med. Lab. (Wright Patherson Air Force Base), Ohio, 1947, Army Air Force Material Command. Engin. Div., Rep. TSE-AA-694-12.

Flanagan, J.C. The critical incident technique. Psych. Bull. 51 (1954) 327-58.

Frusch, E.P.R., J.A. Landeweerd & D.P. Rookmaker. Analyse van "bijna-ongevallen" op overwegen. Eindhoven. Tech. Hogesch. 1971.

Güttinger, V.A. Conflict observation techniques in traffic situations. In: Proceedings first workshop on traffic conflicts. Oslo, Institute of Transport Economics, 1977, pp. 16-21.

Güttinger, V.A. The validation of a conflict observation technique for child pedestrians in residential areas. In: Older, S.J. & J. Shippey (eds.). Proceedings of the 2nd International Traffic Conflicts Technique Workshop, May 1979. Crowthorne, Berkshire, Transp. and Road Res. Lab. 1980, pp. 102-6. TRRL Suppl. Rep. 557.

Güttinger, V.A. From accidents to conflicts; alternative safety measurements. In: Kraay, J.H. (ed.). Proceedings of the third international workshop on traffic conflicts techniques, organised by the international commitee on traffic conflicts techniques ICTCT. Leidschendam, Institute for Road Safety Research, 1982, Resp. R-82-27.

Hayward, J.C. Near miss determination through use of a scale of danger: paper presented at the 51st Annual Meeting of the Highway Res. Road, 1972. Pennsylvania, Transport & Traffic Saf. Center/Pennsylvania State Univ. 1972.

Horst, A.R.A. van der & R.M.M. Symonsma. Behavioural study by the institute for perception IZF-TNO. In: Older, S.J. & J. Shippey (eds.). Proceedings of the 2nd International Traffic Conflicts Technique Workshop, May 1979. Crowthorne, Berkshire, Transp. and Road Res. Lab. 1980, pp. 102-6. TRRL Suppl. Rep. 557.

Paymans, P.J. "Is een bijna-ongeval bijna een ongeval"; een exploratieve analyse van de bijna-ongevallen op overwegen. Amsterdam. Psych. Lab., Vakgroep Funktieleer/ Ergonom. Werkgr. NS, 1972.

Perkins, S.R. & J.I. Harris. Traffic conflict characteristics ; accident potential at intersections. Werren (Mich.), Electr. Mech. Dept. Res. Labs. Gen. Motors Corp., 1967 (Res. Publ. GMR-718).

Proceedings first Workshop on Traffic Conflicts. Oslo, Institute of Transport Economics, 1977.

Spicer, B.R. A pilot study of traffic conflicts at a rural dual carriage-way intersection. Crowthorne (Berkshire), Road Res. Lab./Road User Characteristics Section, 1971 (RRL Rep. LR 410).

Spicer, B.R. A. traffic conflict study at an intersection on the andoversford bypass. Crowthorne (Berkshire), Transp. Road Res. Lab./Dept. Environm., 1972 (TRRL, Rep. LR 520).

Spicer, B.R. A study of traffic conflicts at six intersections. Crowthorne (Berkshire), Transp. Road. Res. Lab./Dept. Environm., 1973 (TRRL, Rep. LR 551).

APPENDIX

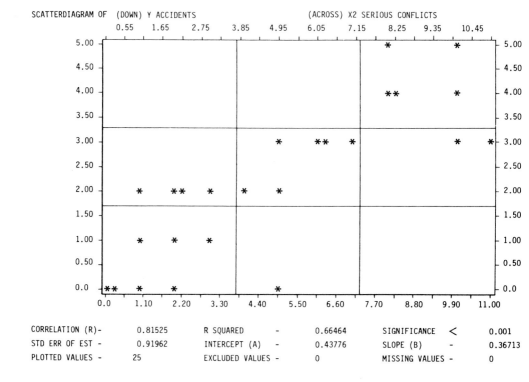

SCATTERDIAGRAM OF (DOWN) Y ACCIDENTS   (ACROSS) X2 SERIOUS CONFLICTS

CORRELATION (R)-	0.81525	R SQUARED	-	0.66464	SIGNIFICANCE	<	0.001
STD ERR OF EST -	0.91962	INTERCEPT (A)	-	0.43776	SLOPE (B)	-	0.36713
PLOTTED VALUES -	25	EXCLUDED VALUES -		0	MISSING VALUES -		0

TRAFFIC CONFLICTS IN BRITAIN: THE PAST AND THE FUTURE

G. B. Grayson
Transport & Road Research Laboratory
Crowthorne, Berkshire RG11 6AU

1.   Introduction

This paper will attempt to provide a brief overview of the traffic conflicts technique in Britain, to assess its past, and to consider how it might develop in the future.   Another paper later in the proceedings will deal with the practical and the detailed aspects of the British technique; the present discussion is more a personal view from one who has only recently become directly involved in conflict work, but who has long been acquainted with its past.   In the discussion three points will be made:

(i)   that real progress has been made in Britain on several fronts in the development of a conflicts technique;

(ii)   that as a result, a realistic view would be that the validity of conflicts has been satisfactorily established; and

(iii)   that future effects need to be directed to wider implementation of the technique and to more diagnostic applications.

2.   The past

The basic idea of the conflict or near-accident had been long established in the fields of industrial and aviation safety before it was applied systematically in a traffic context in the late 1960's. Indeed, it is intriguing to speculate why such a well understood and obviously relevant procedure should have taken so long to reach traffic safety.   Whatever the reasons, the credit for being the first to develop a systematic observation technique always goes to the General Motors team of Perkins & Harris (1967), although it is undoubtedly the case that many were engaged on similar trains of thought at that time, on both sides of the Atlantic, and not least in Britain.   There, the objective from the outset has been to develop a technique that studies events that occur frequently, can be clearly defined, reliably measured, and are related to traffic accidents.   A subsequent objective has been that the use of the technique should not be confined to central research organisations, but should also be accessible and acceptable to the local authorities, who are responsible for much of the practice of road safety in Britain.

NATO ASI Series, Vol. F5
International Calibration Study of Traffic Conflict Techniques
Edited by E. Asmussen
© Springer-Verlag Berlin Heidelberg 1984

In considering how well these objectives have been achieved, it is best to adopt a chronological approach. The first development was to modify the original General Motors definition of a conflict in order to introduce a grading by severity - an integral part of the British technique. The subjective nature of the technique has been criticised by some authors (eg Glennon & Thorson, 1975; Williams, 1981), but the logic of their argument is not always clear, since it seems to be based on the assumption that subjective data are inherently inferior. It must of course be admitted that objective measures would have many advantages, and we have made considerable efforts to develop a fully automated conflict technique. Although unsuccessful, our minds are not closed on the subject. The real point at issue, however, is that of reliability. If the occurrence of conflict events can be recorded in a manner that is reliable, consistent, and repeatable, then the subjective/objective debate is a fruitless one. It is worth noting that Hayward (1972), who advanced one of the first of the objective type definitions, also pointed to the need to train observers to record conflicts, as this in his view was the only economic method of large scale data collection.

The next development might best be described as one in flexibility, in that we have been engaged in a continual search for a better understanding of the factors that influence the relationships between conflicts and accidents. As well as carrying out our own work we have also monitored the international literature. Over time an encouraging consistency has been discernible in experience with, for example, signalised versus non-signalised junctions and with manoeuvre types such as rear end conflicts. If open-mindedness can be regarded as an achievement then it is one that we would claim.

The question of reliability has rightly been of major concern to conflict researchers for many years. We have looked at several different aspects of this problem. For instance, we have established that the results obtained from different observers recording conflicts on the same occasion, and from observers recording the same conflicts on different occasions are both highly correlated. This demonstration of acceptable levels of inter- and intra-observer reliability used research workers as subjects, but we were also interested to find out whether similar results could be obtained from naive subjects, such as might be recruited on a casual basis by local authorities. A contract was therefore given to Nottingham University to develop a training package for use by local authorities, and full scale field trials are taking place this year.

Another aspect of the reliability problem is the question of repeatability, that is, the variation in the numbers of conflicts occurring at any site over a given time. High levels of variability could have serious implications for conflict work, as has been suggested by Glennon & Thorson (1975). A six month study of one site was carried out to investigate this question, and also to provide some indication of the effects on conflict rates of weather, time of year, and day of week. The results (Spicer et al, 1980) indicated that the level of repeatability was more than adequate for both practical and research purposes, and also gave support to the views advanced by Hauer (1978) on the optimal durations for data collection. In the light of British experience, it is impossible to agree with the statement of Williams (1981) to the effect that present conflict methods do not appear to possess the properties of repeatability and consistency.

The most recent development has been to consider the alternatives to conflicts. The possibility of traffic flow as an alternative candidate has been long disposed of; there is now ample evidence that serious conflicts (by our definition) are much more closely related to accidents that is traffic flow. More serious have been the suggestions in recent years from some European workers (eg Malaterre & Muhlrad, 1980) that expert judgements could be as good, or even better than conflicts in diagnosing and predicting hazards. This of course is not a radical idea; for many years traffic engineers have carried out site inspections and assessments, and it may well be that conflicts are no more than the systematic application of properly organised and defined engineering judgements. It is therefore necessary to demonstrate that conflicts can perform better than other procedures in identifying accident risk if one is to be able to argue that conflicts can make a real contribution to the identification of hazardous situations and to the formulation of appropriate countermeasures. Once again, we have been able to do this to our satisfaction.

And so to the question of validity - undoubtedly the central issue in conflict research. It was stated at the outset that we are now effectively convinced of the validity of the traffic conflicts technique, and it is probable that most of those present at this meeting share this conviction to a greater or lesser degree. But what of safety researchers in general, not to mention the practitioners in the field; has the conflects technique been sold to them? The answer very largely must be no. Although research reports on conflict work are circulated among interested parties, papers and articles in the general literature are sadly lacking. Because it was one of the few examples, the paper by Williams (1981) has been regularly quoted in recent times, which is unfortunate since it gives a critical - and partial - review of conflict work. Given the degree of commitment to conflict research that undoubtedly exists in many quarters it is perhaps surprising that no refutation or rejoinder has been made to Williams. It is worth remembering that the more a criticism remains unanswered, the more it gains in its credibility to the outsider. The point needs to be made that safety researchers and practitioners will not come looking for evidence; they need to be convinced by those who are already convinced themselves.

Our approach to the question of validity has been based on small samples by some standards, but these samples have been studied in great detail. The enthusiasm over the earlier results (eg Spicer, 1973) was understandable, given that they were among the first European conflict studies to demonstrate a good relationship between conflicts and accidents. Perhaps inevitably, these results were criticised by the sceptical Glennon & Thorson (1975), who uncharitably and unreasonably termed them 'unbelievable'. A more sensible criticism might have been directed towards the statistical tests used. It would have to be admitted that the Spearman's rho test used at the time was not perhaps the most rigorous in statistical terms, but it was dictated by the smal sample size, and more important, it did accord quite closely with the way in which practitioners operate in real life. To the hard-pressed local authority engineer in the field, the direction of change can be just as meaningful as the estimated size of change.

The whole concept of validity is an exceedingly complex one that cannot be discussed in any real detail here. Looking at the past history of conflict work one can reduce the issue to two alternative criteria for validity. The first is that conflicts should be able to act as predictors for accidents, as surrogate measures that have the same characteristics but occur more frequently. Every worker who has thought at all deeply about the nature of conflicts will have realised that this strong criterion of predictive validity is not appropriate for a variety of reasons, both theoretical and empirical. The second, and far more reasonable criterion is that conflicts should be related to accidents in an orderly and meaningful way. The perfect surrogate measure is an illusion; Hauer (1980) has gone further and argued that correlation is not the best measure of validity, and that instead one should be concerned with obtaining reliable estimates of conflict to accident ratios. We would agree with his views in principle, but recognise that his counsels do depend on the availability of very large data bases.

The British evidence to support the relationship between conflicts and accidents has been presented at earlier international workshops (Older & Shippey, 1977; Baguley, 1982), and the most recent data will be reviewed later in these proceedings. It should be made clear, however, that our position regarding validity has inevitably, and indeed deliberately been influenced by the experience of researchers in other countries as well as by our own results. If one takes an impartial and balanced view of these combined experiences, then we believe there is now enough evidence to convince a reasonable man of the validity of traffic conflicts. Despite Hauer's advice, most researchers have examined the correlations between conflicts and accidents, and the results in general have been far from discouraging once it is appreciated that the conflict-accident ratio must vary with definition, with location, with manoeuvre, and so on. There are many results in the literature that may seem difficult to reconcile, but if approached in a positive and constructive frame of mind there is a remarkable and very encouraging degree of consistency to be found in the results of the work in the last decade. After looking at the evidence it becomes hard to argue with the conclusion that conflicts are related to accidents, if not necessarily predictors of them. Those who have backgrounds in the behavioural sciences will be all too familiar with debates of this nature, but it must surely be difficult for any but the most confirmed sceptic to agree with Williams (1981) when he maintains that 'studies devoted to the evaluation of the traffic conflicts technique have failed to establish that conflicts are related to road accidents.'

Accident statistics, however, should not be seen as the only criterion for the validity of conflict techniques, for it is quite reasonable to argue that they can also be used as measures of operational efficiency. For example, if a traffic engineer plans a new intersection he should have some cause for concern if a high conflict rate is subsequently experienced at that intersection. There need be no necessary implication that the accident rate would increase; it should be sufficient that the very existence of a high conflict rate should suggest that the design of the intersection has been inadequate in some way.

A third way in which conflicts might be used in a 'valid' manner
is in the rather nebulous and ill defined area that has been variously
referred to as quality of life, environmental standards, or even levels
of service. Perhaps the simplest way of describing this idea - and it
is one that is becoming increasingly discussed - is in terms of improvir
customer satisfaction. However, for the argument to have conviction it
is necessary that the customers, ie the users of the road system,
recognise conflicts as undesirable events. We have attempted to gain
some clarification of this issue in a recent experiment carried out by
Nottingham University. In this exercise, drivers who had passed through
an intersection were stopped and interviewed about their experiences,
while at the same time a team of observers recorded the traffic conflict
that occurred at the intersection. The results, it must be admitted,
were far from clear cut. Of all drivers interviewed after having passed
through the intersection, one in five maintained that they had
experienced a conflict situation, defined to them as the need to take
avoiding action. However, less than half of these drivers were actually
observed to be involved in a conflict - as defined by the conflict
observation team. These results show an interesting correspondence with
those obtained by Muhlrad (1982) who found that road users rated risk
more on the basis of slight conflicts than on serious ones or on
accidents. More worrying, though, was the finding that only 41% of
those observed in conflict situations actually declared that they had
been involved in one. However, those who said they had experienced a
'conflict' gave the intersection a significantly higher rating on a
scale of riskiness than those drivers who did not record a conflict.

Although this pilot study may well have raised more questions than
it answered, it seems very likely that this approach will become more
common in the future. Subjective assessments of safety are increasingly
coming to be recognised as a potentially important part of both safety
research and safety policy, even if it must be admitted that at present
the research community is still poorly equipped to contribute to the
debate. The methodological problems are formidable, but there seems
little doubt that conflict studies will have an important role to play
in this area, and it is encouraging to find that some countries have
already begun to tackle this problem.

3.   The future

At this point it is appropriate to turn from the past to the future
and ask the age-old question 'quo vadis?' The British view is that
conflict work must follow two avenues. The first of these is

implementation. If conflict techniques are to make any lasting contribution they need to move from research establishments to the practitioners who have to cope with problems in the field. We have made considerable efforts to make the conflict technique more accessible to these users, and will continue to do so. We also hope to monitor the results of local authorities in their application of conflict techniques. This we feel could be valuable not just because longer term studies are sadly lacking in the literature, but also because the results could help in the compilation of what Ezra Hauer has aptly termed the conflicts 'catalogue'. The important message is that we consider conflicts to be a dependable tool that should be put to work in tackling safety problems.

The second approach, like the first, depends on accepting the validity of the conflicts technique, but this time in a research rather than an applied context. Here, the argument is that conflicts should not be regarded simply as unfulfilled accidents. This might seem paradoxical, given the considerable effort that has been devoted in attempting to show that conflicts are correlated with accidents, and might therefore be used as surrogate measures for them. But while the two may be on the same continuum, they make very different contributions to safety research. Accident studies all too often operate like black boxes, where countermeasures are implemented, accident data are collected, statistics are applied, and verdicts are pronounced, where the sole concern has been with the question - 'does the countermeasure work?'

By contrast, because conflicts are a much richer source of data they offer the possibility of gaining more understanding of the road system, and of answering the questions how and why countermeasures work - or do not work. To be able to use conflicts as a diagnostic tool in the identification of hazards and in the formulation of countermeasures has long been the aim of traffic conflict workers. In Britain, conflict techniques are now used regularly in this way by several workers in local authorities, and it is expected that this application will become more widespread in the future. But the potential of the traffic conflicts technique can go further than this. If one accepts their validity, then conflicts can provide a very powerful research tool for the study of road user behaviour. It is interesting to note that this approach has already been followed in Britain by those working in the field of pedestrian conflicts (Howarth & Lightburn, 1980), who have not waited to establish validity and who have obtained valuable data while doing so. In the field of vehicle conflicts the payoff could be equally high; in the future we believe that less emphasis should be placed on conflicts as alternatives to accidents, and more on conflicts as a means of understanding how the traffic system works. Our intentions might be put succinctly in saying that the time has now come to study conflicts rather than just count them. Only in this way can conflict research make a real contribution to a greater understanding of the system that we study, and this, ultimately, is what road safety research should be about.

The position that we have reached does not mean that we feel there is not need for further international contact; indeed the opposite is the case. We acknowledge the contribution of others in the past, and recognise that our present position has only been achieved by the effort of many. Williams (1981) maintained that 'the conflicts technique has gained acceptance among practitioners who appear to be unaware of the uncertainty of evaluation results.' This quotation shows clearly that he is not familiar with conflict workers, who as a group show more concern, heart-searching, self-criticism and doubt than any other group of safety researchers. (Grayson's Law: the level of uncertainty is exponentially related to the number of conflict researchers in the room

But at the same time it is doubtful if any other group of safety workers has the same degree of interest and commitment to learn from and with others. Traffic conflicts have become one of the truly international areas of research, which is why this Malmo exercise has been so well supported. At the same time, though, it is worth looking back to the conclusion of the Paris Workshop, when McDowell (1980) urged that conflicts should be sold more widely. Now that conflict research is more than 15 years old, the time should have come to convince others of what has been accomplished in that period. One hopes that the Malmo study will provide an incentive for the wider dissemination of the results of conflict research.

References

BAGULEY, C J, 1982. The British traffic conflict technique: state of the art report. Proceedings of the Third International Workshop on Traffic Conflicts Techniques. Institute for Road Safety Research SWOV, Leidschendam.

GLENNON, J C and THORSON, B A, 1975. Evaluation of the traffic conflicts technique. Midwest Research Institute, Kansas City.

HAUER, E, 1978. Traffic conflict surveys: some study design consideration TRRL Report SR352. Transport and Road Research Laboratory, Crowthorne.

HAUER, E, 1980. Methodological assessment of the techniques. Proceedings of the Second International Traffic Conflicts Technique Workshop. TRRL Report SR557. Transport and Road Research Laboratory, Crowthorne.

HAYWARD, J C, 1972. Near-miss determination through use of a scale of danger. Highway Research Record No 384, 24-34. Highway Research Board, Washington DC.

HOWARTH, C I and LIGHTBURN, A, 1980. How drivers respond to pedestrians and vice versa. In: D J OBORNE AND J A LEVIS (Eds) 'Human Factors in Transport Research', Vol II. Academic Press, London.

MALATERRE, G and MUHLRAD, N, 1980. Conflicts and accidents as tools for a safety diagnosis. Proceedings of the Second International Traffic Conflicts Technique Workshop. TRRL Report SR557. Transport and Road Research Laboratory, Crowthorne.

McDOWELL, M R C, 1980. A personal overview. Proceedings of the Second International Traffic Conflicts Technique Workshop. TRRL Report SR557. Transport and Road Research Laboratory, Crowthorne.

MUHLRAD, N, 1982. The French traffic conflict technique: a state of the art report. Proceedings of the Third International Workshop on Traffic Conflict Techniques. Institute for Road Safety Research SWOV, Leidschendam.

OLDER, S J and SHIPPEY, J, 1977. Traffic conflict studies in the United Kingdom. Proceedings, First Workshop on Traffic Conflicts. Institute of Transport Economics, Oslo.

PERKINS, S R and HARRIS, J I, 1967. Traffic conflict characteristics: accident potential at intersection. Research Publication GMR-718. General Motors Corporation, Warren, Michigan.

SPICER, B R, 1973. A study of traffic conflicts at six intersections. TRRL Report LR551. Transport and Road Research Laboratory, Crowthorne.

SPICER, B R, WHEELER, A H and OLDER, S J, 1980. Variation in vehicle conflicts at a T-junction and comparison with recorded collisions. TRRL Report SR545. Transport and Road Research Laboratory, Crowthorne.

WILLIAMS, M J, 1981. Validity of the traffic conflicts technique. Accident Analysis & Prevention 13, 133-145.

THE DEVELOPMENT AND USAGE OF TRAFFIC CONFLICT TECHNIQUE ON THE SWEDISH NATIONAL ROAD NETWORK

Mats-Ove Mattsson
Section for Development
Swedish National Road Administration
S 781 87 Borlänge, Sweden

1. Background

When measures were taken on sections of road within the Swedish national road net-work up until approximately the mid-1970's, aspects concerning traffic safety were not of predominant concern. This meant that problems existing with respect to traffic safety were often attended to in connection with either the performance of larger investment projects (like bypasses), or in connection with less extensive improvement works (such as the easing of bends on curving roads). Traffic pass-ability, vehicle economy, employment etc. were also factors in addition to the traffic safety aspect constituting the grounds for the carrying out of such projects.

Towards the end of the 1970's, a recommendation was made by the Swedish government in a bill on traffic, that the Swedish National Road Administration (SNRA) increase its efforts in the area of traffic safety. This meant in practice that the SNRA was to invest greater resources in certain demarcated places with traffic safety problems (blackspots).

Certain blackspots were chosen at that point in time. The selection was made on subjective grounds, based on experience. In order to obtain more uniform methods for judging and analysing local traffic safety problems, a model was worked out at the head office of the SNRA. This model, the Regional Road Administration Work on Traffic Safety (VFTSA), was published as a report in 1978 (4) and describes in part the following:

- methods for pointing out blackspots through systematically analysing accidents reported by the police on a specific road network
- methods for recognizing and obtaining information on locations considered danger-ous from a traffic point of view, but where no accidents have been reported by the police
- methods for analysing these locations pointed out in order to find the measures suitable for attending to the problem
- methods for establishing priorities on the locations analysed with respect to the gain in traffic safety, the cost, etc.

NATO ASI Series, Vol. F5
International Calibration Study of Traffic Conflict Techniques
Edited by E. Asmussen
© Springer-Verlag Berlin Heidelberg 1984

Generally, there is no difficulty in locating places with traffic safety problems using the VFTSA model. The difficulties arise when making an analysis of the blackspots in order to find a suitable solution. If the number of accidents registered at the locations pointed out is too low, which often is the case, it is extremely difficult to choose relevant measures for the improvement of traffic safety. Problems arise when setting priorities also. Priorities are given according to the estimated gain in traffic safety. This gain is directly dependent on the number of accidents which have occurred, and for which measures taken had supposedly prevented.

A larger accident base is therefore required in order to be able to deal with these problems. Statistically, accidents occur infrequently. It is also difficult to noticeably improve the routines established with the police with respect to the reporting of accidents. One remaining possibility then, is to supplement the accident material with registered events and behaviour in traffic which occur more frequently than accidents, but which have a direct relation to them. This constitutes the background as to why we at the Swedish National Road Administration consider it important to use conflict studies when working on traffic safety.

2. Adaptation of the existant traffic conflicts technique

A traffic conflicts technique was developed in Sweden during the 1970's at the Lund Institute of Technology. This technique was, however, developed primarily with reference to urban environments where the velocity is generally low. While urban sections do occur within the Swedish national road network, the overwhelmingly predominant type of road environment is rural. Because of this fact, we started a project in 1978 at the Lund Institute of Technology for the purpose of adapting the traffic conflicts technique already in existence to useage in rural conditions.

During the first stage of the project, the existent urban technique was considered to be suitable for rural conditions as well, but with one reservation. This was that the fixed limit value used to judge the degree of seriousness of a conflict, 1.5 seconds, be changed to a value dependent on the velocity. The figure on the following page indicates how this limit value varies with the velocity.

THE LIMIT VALUE BETWEEN SERIOUS AND NON-SERIOUS
CONFLICTS AND ITS VARIATION WITH VELOCITY AND
THE TO – VALUE

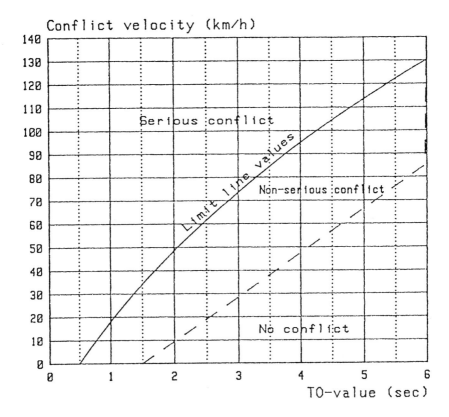

TO – VALUE: This is the time in seconds which would have elapsed from the exact
time one of two roadusers in a conflict situation initiated either a
deceleration or a veering motion, and the time the two parties involved
had reached the envisioned point of collision had both of them continued at the same velocity and on the same course.

A more detailed description of the urban technique is contained in reference no. 1
and in the report of the Swedish technique.

Moreover, it can be mentionned that the relation between conflicts and accidents
indicated in the urban technique was not generally considered to be representative
for the national road network as a whole.

One prerequisite for the continued useage of the traffic conflicts technique was
that personnel at the regional level of the Swedish National Road Administration

could have the possibility of taking advantage of it. Stage two was thus directed
at producing a training program for personnel at the Regional Road Administrations
throughout the country. An instruction manual in conflict studies was therefore
drawn up based on the experiences gained from courses completed, conflict studies
performed etc. The manual was primarily to be of use during the training program,
as well as in connection with the independent performance of conflict studies.
Furthermore, it can be regarded as an information publication in the field of
traffic conflicts technique.

3. The current situation

At present, there is one observer, on the average, trained in the field of traffic
conflicts studies in every Regional Road Administration throughout the country.
Every Regional Road Administration is thus in the position of being able to start
its own conflict study within the near future. The following can facilitate the
start of such studies:

- the instruction manual mentionned previously
- advisory consultation from both the head office of the Swedish National Road
  Administration and from the Lund Institute of Technology
- a film containing information on traffic conflicts technique and which also
  has examples from conflict situations as exercises

The Head Office of the Swedish National Road Administration has not made any stip-
ulations as to the useage of the traffic conflicts technique. Each Regional Road
Administration can independently make its own decision from case to case as to the
performance of a prospective conflict study.

Examples of some conflict studies in which the Swedish National Road Administration
is currently involved are:

- the Tånga junction in Falkenberg. This involves a before and after study in order
  to measure the effect of the installation of traffic lights.
- Sölvesborgs Road in Karlshamn. This entails a study for which the purpose was
  to arrive at a solution to the traffic safety problems arising for bicyclists
  and mopedists.

4. Future prospects

It can be taken for granted that the traffic conflicts technique will be a factor
in the work on traffic safety within the Swedish National Road Administration in
the future. It can be assumed that the Regional Road Administrations will be able
to carry out an increased extent of independent conflict studies in the very near

future as more and more personnel are trained. This will most probably mean that a type of routine will be established; i.e., places will be selected at regular intervals which are well suited for conflict studies and where the need for conflict studies exists. Video films will be made at certain places selected for the purpose of effectively illustrating the conflicts registered. It can also be presumed that the various Regional Road Administrations will carry out an increased number of conflict studies after taking measures to solve a problem in order to follow up the effect. If the measure proves to be a poor solution according to the follow-up, other measures can be tried and subsequently evaluated.

Moreover, we hope that in the future we will be able to compile the various conflict studies completed to find a correlation between the different types of conflicts and types of accidents. This correlation can then subsequently be broken down into various groups of junction layout-types.

5. Summary and conclusions

In summary, the following conclusions can be drawn on the traffic conflicts technique and our utilization of it.

- The traffic conflicts technique gives a uniform method whereby conflicts can be registered. This implies that more comprehensive compilations and evaluations can be made irrespective of the geographical location and the person or persons responsible for the investigation.
- The traffic conflict technique is grounded on the assumption that conflicts are related to accidents. This implies that conflict studies are applicable in connection to work on traffic safety.
- Conflict data is primarily used as a supplement to accident data in order to enhance the possibility of selecting the correct measure to improve traffic safety. Furthermore, through the performance of before and after studies, a rough appreciation can be obtained from a traffic safety point of view, of the effect of measures taken.
- The traffic conflicts technique is best suited at locations with a limited geographical area where there is a large enough traffic flow that a sufficient number of conflicts can be registered within a reasonable length of observation time and without using too many resources.
- At present, we do not have extensive material or experience from conflict studies carried out on our road network. It is our hope in the future that we will be able to find a correlation between conflicts and accidents when we have more data compiled. This, in addition to the experience gained, can lead to an improvement of the present conflict model.

We deem it urgent to start conflict studies at the Regional Road Administrations as

soon as possible in order to improve the future work on traffic safety and to
increase our knowledge and experience as a result of the conflict studies
performed.

References

1. Hydén, C., "En konfliktteknik för riskbestämning i trafiken" (A traffic
   conflicts technique for determining risk). Lund Institute of Technology,
   Lund, 1976.

2. Gårder, P., "Konfliktstudier i landsvägskorsningar" (Conflict studies at
   rural junction). Lund Institute of Technology, Lund, 1982.

3. Mattsson,M-O., "Utbildningsmanual i konfliktstudier" (Instruction manual
   for conflict studies). Swedish National Road Administration, Borlänge, 1983.

4. Åkerlund, O., "VF-Trafiksäkerhetsarbete" (Regional Road Administration Work
   on Traffic Safety). Swedish National Road Administration, Stockholm, 1978.

THE TRAFFIC CONFLICT TECHNIQUE OF THE UNITED STATES OF AMERICA

James Migletz and William D. Glauz
Midwest Research Institute
Kansas City, Missouri 64110

The traffic conflict technique (TCT) has been practiced in the United States for over 15 years. Traffic conflict observation began on a large scale in 1969 when the U.S. Federal Highway Administration (FHWA) awarded contracts to three state highway agencies. Teams of observers were employed and the TCT was practiced on a regular basis. Other states also began applying the TCT. The TCT was being utilized to solve operational problems at intersections. There also was the belief that an intersection safety record could be determined without relying on historical accident records. However, because there was a lack of a proven, direct relationship between accidents and conflicts, the United States TCT has received less emphasis from highway administrators and is not being practiced on the large scale that it once was.

The international traffic conflict calibration study provides the opportunity to present the United States TCT so that it may be discussed and compared with the techniques practiced in other countries. Results of this international effort will help achieve the goal of improving the safety and efficiency of our streets and highways.

The Midwest Research Institute (MRI) has been involved in two studies involving traffic conflict research. The first study, completed in September 1979, was National Cooperative Highway Research Program (NCHRP) Project 17-3 and is reported in NCHRP Report 219:[1] "Application of Traffic Conflict Analysis at Intersections." The objective of this research was to develop a standardized set of definitions and procedures that would provide a cost-effective method for measuring traffic conflicts.

MRI is currently conducting a research project for the FHWA entitled: "Identification and Quantification of Relationships Between Traffic Conflicts and Accidents." The objectives of this research are:

1. To quantify the relationships between specific types of traffic conflicts and analogous accident types for specific intersection conditions.

2. To identify the expected and abnormal conflict rates by determination of means and variances of conflict types.

NATO ASI Series, Vol. F5
International Calibration Study of Traffic Conflict Techniques
Edited by E. Asmussen
© Springer-Verlag Berlin Heidelberg 1984

The current research is the calibration of the United States TCT.

The remainder of the paper describes the United States TCT and is organized as follows:

A. Definition
B. Conflict Types and Road Situations
C. Data Collection Procedure
D. Data Treatment
E. Training Procedure
F. Observation Periods
G. Evaluation
H. Malmö Experiment
I. References

A. Definitions

This section presents traffic conflict definitions and definitions needed to utilize the data collection form. It begins with a generalized or global definition-- a framework into which most operational definitions may be placed. Then 13 basic types of intersection conflict situations are described.

The generalized definition of a traffic conflict which will form the basis for specific operational definitions is presented first:

> A traffic conflict is a traffic event involving two or
> more road users, in which one user performs some atypical or
> unusual action, such as a change in direction or speed, that
> places another user in jeopardy of a collision unless an eva-
> sive maneuver is undertaken.

Generally, speaking, the road users are motor vehicles, but the definition is broader in that they could also be pedestrians or cyclists.

The action of the first user is atypical or unusual in that it is not an action that every road user, or the typical road user, would perform under the same circumstances, although it need not necessarily be an infrequent or extreme action. An example might be precautionary braking by a motorist driving through an intersection, even though there is no cross traffic. This restriction does, however, rule out certain types of movements that all (or nearly all) users initiate under the same conditions. Examples here are stopping for a stop sign or red traffic signal indication, or reducing speed to negotiate a turn in the roadway. Thus, the definition implicitly excludes actions that are in compliance with a traffic control device or that are required in response to the roadway geometrics.

Within the context of this general definition, which is conceptual in nature, it is not necessary that there actually be an evasive maneuver or that there actually be an impending collision. It suffices that the instigating action or maneuver threatens another user with the possibility of a collision and, thereby, places the user in the position of probably taking some evasive maneuver. Clearly, however, many collisions occur in which there are no evasive maneuvers; they would be included as extreme cases under this broad definition. Further, there are often "near miss" situations in which a second driver either is unaware of a collision potential, or is unusually adept at estimating time intervals and clearances, and chooses not to make an evasive maneuver; these situations are also included under the broad definition.

To further clarify this general definition, counter examples may be given. For example, the definition would exclude "evasive maneuvers" that are strictly precautionary in nature. For example, it would not include braking or swerving (lane changing) of a through vehicle in response to an anticipated opposing left turn, perhaps instigated by the opposing driver turning his wheels (but not encroaching on the lane of the through vehicle). Likewise, it would not include braking or swerving occasioned by the presence of a stopped vehicle on a cross street, which may "threaten" to encroach but does not actually do so. Another general class of exclusions is violations such as "run red light" and "run stop sign," unless such violations occur in the presence of a through-vehicle that is placed in jeopardy of a collision.

Adopting the general definition as a basis for practical operational definitions requires that certain assumptions be made. If the TCT is to be implemented widely in the United States, any operational definitions must avoid or minimize the use of sophisticated equipment and painstaking measurements, whether in the field or later in the office. Thus, the operational definitions must be suitable for application by human observers. Moreover, it is unlikely that such definitions would be used frequently if they required highly educated and experienced traffic engineers as observers; the definitions should be amenable to use by persons such as traffic technicians, with suitable training.

With these costraints, it is obvious that the operational definitions must encompass readily observable events. It was judged highly unlikely, on the basis of the research and operational experiences throughout the world in the last 10 years, that relatively unskilled persons could be trained to consistently "observe" traffic events, even of a near-miss variety, unless some reaction to the event was elicited in one of the road users. Thus, the general approach taken by General Motors[2] was adopted. The traffic event must elicit an evasive maneuver (braking or swerving) by the offended driver.

An intersection traffic conflict can then be described, operationally, as a traffic event involving several distinct stages:

1. One vehicle makes some sort of unusual, atypical, or unexpected maneuver.
2. A second vehicle is placed in jeopardy of a collision.
3. The second vehicle reacts by braking or swerving.
4. The second vehicle then continues to proceed through the intersection.

The last stage is necessary to convince the observer that the second vehicle was, indeed, responding to the offending maneuver and not, for example, to a traffic control device.

Within this framework a basic set of operational definitions can be stated, corresponding to different types of instigating maneuvers. The 14 intersection conflict situations that appear to be potentially useful in pinpointing operational or safety deficiencies are presented below:

1. Left Turn, Same Direction
2. Slow Vehicle, Same Direction
3. Lane Change
4. Right Turn, Same Direction
5. Opposing Left Turn
6. Left Turn, Cross Traffic from Left
7. Thru Cross Traffic from Left
8. Right Turn, Cross Traffic from Left
9. Left Turn, Cross Traffic from Right
10. Thru Cross Traffic from Right
11. Right Turn, Cross Traffic from Right
12. Opposing Right Turn on Red
13. Pedestrian
14. Secondary

The following paragraphs describe each one. Note that all are described from the viewpoint (direction of travel) of a driver that is being offended or conflicted with, rather than that of the road user instigating the conflict situation. The corresponding diagrams of conflict situations are presented in Figure 1.

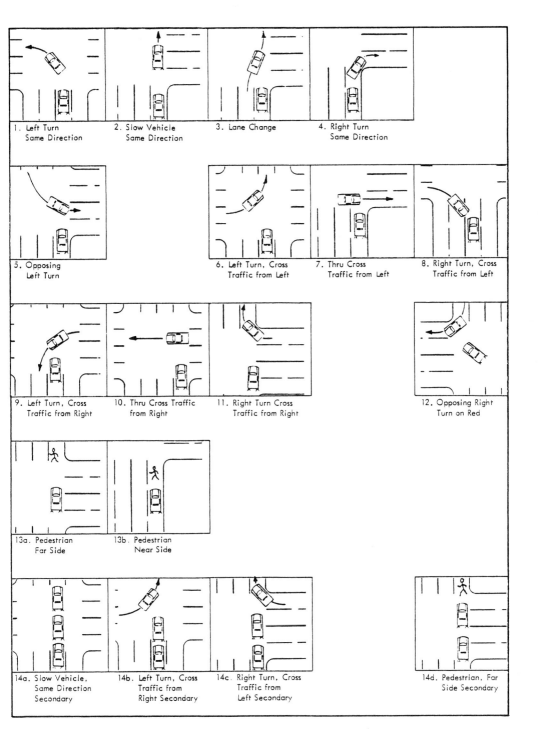

Figure 1 - Intersection Conflict Situations

1.   <u>Left turn, same direction</u> - A left-turn, same-direction conflict situation occurs when an instigating vehicle slows to make a left turn, thus placing a following, conflicted vehicle in jeopardy of a rear-end collision. The conflicted vehicle brakes or swerves, then continues through the intersection.

2.   <u>Slow vehicle, same direction</u> - A slow-vehicle, same-direction conflict situation occurs when an instigating vehicle slows while approaching or passing through an intersection, thus placing a following vehicle in jeopardy of a rear-end collision. The following vehicle brakes or swerves, then continues through the intersection.

The reason for the vehicle's slowness may or may not be evident, but it could simply be a precautionary action or as the result of some congestion or other cause beyond the intersection. If, however, a vehicle slows while approaching or passing through an intersection because of another vehicle that is approaching or within the intersection, the slowing vehicle is itself a conflict vehicle responding to some other conflict situation. In this case, a vehicle following the slowing vehicle is said to face not a slow-vehicle, same-direction situation, but rather a secondary conflict situation, which is described subsequently.

3.   <u>Lane change</u> - A lane-change conflict situation occurs when an instigating vehicle changes from one lane to another, thus placing a following, conflicted vehicle in the new lane in jeopardy of rear-end sideswipe collision. The conflicted vehicle brakes or swerves, then continues through the intersection. However, if the lane change is made by a vehicle because it is in jeopardy, itself, of a rear-end collision with another vehicle, the following vehicle in the new lane is said to be faced not with a lane-change conflict situation, but with a secondary conflict situation, which is described subsequently.

4.   <u>Right turn, same direction</u> - A right-turn, same-direction conflict situation occurs when an instigating vehicle slows to make a right turn, thus placing a following, conflicted vehicle in jeopardy of a rear-end collision. The conflicted vehicle brakes or swerves, then continues through the intersection.

By convention, in the following conflict situations, the conflicted vehicle is presumed to have the right-of-way, and this right-of-way is threatened by some other road user. Situations such as when a "conflicted" vehicle is in jeopardy of a collision because it is running a red light, for example, are not treated as traffic conflicts.

5.   <u>Opposing left turn</u> - An opposing left-turn conflict situation occurs when an oncoming vehicle makes a left turn, thus placing the conflicted vehicle in jeopardy

of a head-on or broadside collision. The conflicted vehicle brakes or swerves, then continues through the intersection.

6. <u>Left turn, cross traffic from left</u> - A left-turn, cross-traffice-from-left conflict situation occurs when an instigating vehicle approaching from the left makes a left turn, thus placing a conflicted vehicle in jeopardy of a broadside or rear-end collision. The conflicted vehicle brakes or swerves, then continues through the intersection.

7. <u>Thru cross traffic from left</u> - A thru, cross-traffic-from-left conflict situation occurs when an instigating vehicle approaching from the left crosses in front of a conflicted vehicle, thus placing it in jeopardy of a broadside collision. The conflicted vehicle brakes or swerves, then continues through the intersection.

8. <u>Right turn, cross traffic from left</u> - The right-turn, cross-traffic-from-left conflict situation is rather unusual. It occurs when an instigating vehicle approaching from the left makes a right turn across the center of the roadway and into an opposing lane, thus placing a conflicted vehicle in that lane in jeopardy of a head-on collision. The conflicted vehicle brakes or swerves, then continues through the intersection.

9. <u>Left turn, cross traffic from right</u> - A left-turn, cross-traffic-from-right conflict situation occurs when an instigating vehicle approaching from the right makes a left turn, thus placing the conflicted vehicle in jeopardy of a broadside collision. The conflicted vehicle brakes or swerves, then continues through the intersection.

10. <u>Thru cross traffic from right</u> - A thru, cross-traffic-from-right conflict situation occurs when an instigating vehicle approaching from the right crosses in front of the conflicted vehicle, thus placing it in jeopardy of a broadside collision. The conflicted vehicle brakes or swerves, then continues through the intersection.

11. <u>Right turn, cross traffic from right</u> - A right-turn, cross-traffic-from-right situation occurs when an instigating vehicle approaching from the right makes a right turn, thus placing the conflicted vehicle in jeopardy of a broadside or rear-end collision. The conflicted vehicle brakes or swerves, then continues through the intersection.

12. <u>Opposing right turn on red</u> - An opposing right-turn-on-red conflict situation can occur only at a signalized intersection that includes a protected left-turn phase. The situation occurs when an oncoming vehicle makes a right turn on

red during the protected left-turn phase, thus placing a left turninng, conflicted vehicle (which has the right-of-way) in jeopardy of a broadside or rear-end colli- sion. The conflcted vehicle brakes or swerves, then continues the left-turn move- ment through the intersection.

13. Pedestrian - A pedestrian conflict situation occurs when a pedestrian (the instigating road user) crosses in front of a vehicle that has the right-of- way, thus creating a potential collision situation. The vehicle brakes or swerves, then continues through the intersection. Any such crossing on the near side or far side of the intersection is liable to be a conflict situation. However, pedes- trian movements on the right and left sides of the intersection are not considered liable to create conflict situations if such movements have the right-of-way, such as during a "walk" phase.

14. Secondary conflict situations - In any of the foregoing 13 conflict situa- tions, it is possible that when the conflicted vehicle makes an evasive maneuver, it places yet another road user in jeopardy of a collision. This type of traffic event is called a secondary conflict situation (it is comparable to the GM-defined "previous conflict"). Nearly always, the secondary conflict situation will appear the same as a slow-vehicle, same-direction conflict situation or a lane-change situ- ation, as described previously. The difference, of course, is that the secondary conflict situation is the result of an instigating vehicle that slowed or swerved (changed lanes)in response to some other conflict situation.

In all of the foregoing operational definitions it is necessary that the conflicted vehicle, the one that is placed in jeopardy of a collision, actually takes an eva- sive maneuver, as evidenced by obvious braking or swerving. In most cases the brak- ing will be observed as brake-light indications, although a noticeable "diving" of the vehicle in the absence of brake lights is also acceptable evidence of an evasive maneuver.

Alternate operational definitions were also used in the field tests of previous research to determine their value relative to the baseline definition. For each of the 13 conflict situations (14, including secondary conflicts), one less re- strictive and one more restrictive definition were examined. These definitions were developed on the basis of observations of actual practice and on evidence in the literature.

Definitions tested included paired-vehicle conflicts, that require vehicles to be traveling as a pair; opportunities which could become conflicts only if a conflicted vehicle is relatively close and reacts by braking and swerving; and severe conflicts as defined by Hydén,[3] with a time-to-collision threshold of 1.5 sec, as determined subjectively by trained observers. The three definitions were not utilized in the current traffic conflict research.

B. Conflict Types and Road Situations

The operational definitions are such that traffic conflicts involving road users, including motor vehicles, cyclists, and pedestrians, can be recorded. However, past and current research conducted by MRI has focused on conflicts involving motor vehicles. Test sites had been selected for the prime purpose of observing vehicle-vehicle interactions. Intersections near the center of cities which contain large amounts of pedestrian traffic were not selected because the same intersections also contain higher traffic volumes and are subject to congested traffic flow. Conflict studies were conducted during the summer months when schools were on vacation. Any intersections near schools would, therefore, not carry large amounts of pedestrian of cyclist flow.

The United States TCT can be applied at both signalized and unsignalized intersections. Studies have been conducted at intersections comprised of four approaches (legs) and three approaches. However, the technique can be readily adapted to other intersection configurations.

C. Data Collection Procedure

A traffic conflict study is performed by human observers positioned along the roadway to be able to see conflicts relative to the observer's approach. The observer utilizes a count board to tally conflicts. At the end of an observation period, the conflict counts are recorded on the data collection form shown in Figure 2.

The following is a description of the definitions and codes on data collection forms that are utilized to uniquely identify a conflict study and describe the conditions under which data were collected.

Loc - The unique two-digit number assigned to each study intersection.

D - The day of week, 1-7 corresponding to Sunday-Saturday, respectively.

Date - The two-digit month and two-digit day.

Obs - The unique two-digit number assigned to each observer.

MRI

INTERSECTION TRAFFIC CONFLICTS

1	2	3	4	5	6	7	8	9
LOC		D	Date				Obs	

1 N, 2, 3, 4

C - Conflict SC - Secondary Conflict

Intersection of (1 - 3) ______ and (2 - 4) ______

COUNT START TIME (MILITARY)	Time Period	Duration	Road Condition	Approach	Left Turn Same Direction 1		Slow Vehicle 2		Lane Change 3		Right Turn Same Direction 4		Opposing Left Turn 5		Left Turn from Left 6		Cross Traffic from Left 7		Right Turn from Left 8		Left Turn from Right 9		Cross Traffic from Right 10		Right Turn from Right 11		Right Turn on Red 12						TOTAL
					C	SC	C	SC	C	SC	C	SC	C	SC	C	SC	C	SC	C	SC	C	SC	C	SC	C	SC	C	SC	C	SC	C	SC	
	10 11	2 3	4	5	6 7	8 9	20 1	2 3	4 5	6 7	8 9	30 1	2 3	4 5	6 7	8 9	40 1	2 3	4 5	6 7	8 9	50 1	2 3	4 5	6 7	8 9	60 1	2 3					
7:15	01																																
7:45	02																																
8:15	03																																
9:15	04																																
9:45	05																																
10:15	06																																
11:15	07																																
11:45	08																																
12:15	09																																
12:45	10																																
14:15	11																																
14:45	12																																
15:15	13																																
16:15	14																																
16:45	15																																
17:15	16																																
TOTAL																																	

NOTES AND COMMENTS:

Figure 2 - Data Collection Form

Duration - A typical observation period has a duration of 25 min, in which case the duration code is ØØ. An observation period of any other duration would be recorded in minutes.

Road condition - The pavement surface condition during each observation period. Observation under dry road condition is coded as a Ø. Observation under wet road condition is coded as a 1.

Approach - The leg of the intersection being observed. The approaches are numbered 1, 2, 3, and 4 and correspond to the compass directions north, east, south, and west, respectively.

To start on time, the observer will have to arrive at the test site 15 to 30 min before starting to count. A very important preliminary activity is to watch the traffic, and become familiar with the major traffic movements, the signalization characteristics, and any unusual activities. Also, locations of nearby driveways, parked vehicles, or other features that may cause traffic problems should be noted.

Observation points are selected next. For conflicts observation, a location from 100 to 300 ft (30 to 91 m) upstream of the intersection and on the right side of the approach is usually best. This depends on vehicle speeds and approach geometry. At high-speed locations, a location should be picked farther away from the intersection so that all actions and maneuvers relating to the intersection can be observed.

The observer's vehicle should be parked off the roadway. If an adequate parking place is not available, the observer will have to perform the study outside of the vehicle.

Once the observation positions are determined, the data collection forms should be prepared. All heading information should be completed and double checked before any data are collected. The count board has to be "zeroed." If there is more than one observer, watches will have to synchronized.

Slightly different procedures are used during conflict observation of signalized and unsignalized intersections. At signalized intersections, the right-of-way alternates from one street to the other by means of signal timing. This means that while a particular street has the right-of-way (green light), conflicts will be occurring relative to this street. When the signal changes, conflicts will occur relative to the other street (by definition, conflicts occur relative to the street that has the right-of-way). At unsignalized (2-way stop-controlled) intersections, one street has the right-of-way all of the time. Therefore, conflicts will always occur relative to that street.

At typical 4-leg signalized intersections, conflicts will occur relative to all four approaches, while at typical 4-leg unsignalized (2-way stop) intersections, conflicts will occur relative to the two approaches that always have the right-of-way. Thus, conflicts are observed from four approaches at signalized intersections and from the two approaches with the right-of-way at unsignalized intersections.

A typical observation day to obtain a series of conflict counts is between the hours of 0700-1800 hr. Figure 3 contains the observation schedule for a typical conflict study. The upper portion shows the 16, 30-min observation periods and how the periods are spaced throughout a day. Observation is conducted in groups of 3 or 4 periods as described in the middle portion of the figure. The detailed period-by-period schedule is described in the bottom portion of the figure.

D. Data Treatment

At the end of a conflict study the observers exchange forms to review codes and conflict counts to identify errors. Typical coding errors include incorrect dates and approach numbers. A typical error in recording conflicts occurs when primary conflict counts are recorded in the columns reserved for secondary counts. (There can be no secondary counts without at least an equal number of primary counts.) A supervisor also reviews the conflict forms at the office.

Various levels of sophistication can be used in the analysis of conflict counts. One can automate the process using computers or analyze the data by hand. For the research study, the conflict data of the study intersections were entered into a computerized data base via a computer terminal and a data entry/validity check program. The set of entered data was then checked via a computer program to identify values of secondary conflicts that were greater than the value of the corresponding primary conflicts. In the current research, 1.2% of the data collection periods contained errors of this type. After all discovered errors in the conflict counts were corrected, the data base was manipulated into the desired format to facilitate data reduction and analysis.

E. Training Procedure

Persons who will be conflict observers must be extremely conscientious and trustworthy. They will be on their own much of the time, without supervision. They must be trusted to record what they see, and not to fabricate data.

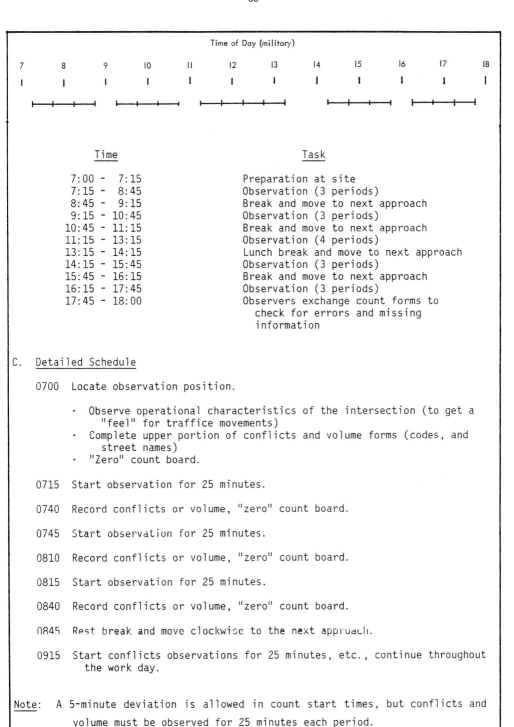

Time of Day (military)

7   8   9   10   11   12   13   14   15   16   17   18

Time	Task
7:00 - 7:15	Preparation at site
7:15 - 8:45	Observation (3 periods)
8:45 - 9:15	Break and move to next approach
9:15 - 10:45	Observation (3 periods)
10:45 - 11:15	Break and move to next approach
11:15 - 13:15	Observation (4 periods)
13:15 - 14:15	Lunch break and move to next approach
14:15 - 15:45	Observation (3 periods)
15:45 - 16:15	Break and move to next approach
16:15 - 17:45	Observation (3 periods)
17:45 - 18:00	Observers exchange count forms to check for errors and missing information

C. Detailed Schedule

0700 Locate observation position.

- Observe operational characteristics of the intersection (to get a "feel" for traffice movements)
- Complete upper portion of conflicts and volume forms (codes, and street names)
- "Zero" count board.

0715 Start observation for 25 minutes.

0740 Record conflicts or volume, "zero" count board.

0745 Start observation for 25 minutes.

0810 Record conflicts or volume, "zero" count board.

0815 Start observation for 25 minutes.

0840 Record conflicts or volume, "zero" count board.

0845 Rest break and move clockwise to the next approach.

0915 Start conflicts observations for 25 minutes, etc., continue throughout the work day.

Note: A 5-minute deviation is allowed in count start times, but conflicts and volume must be observed for 25 minutes each period.

Figure 3 - Daily Conflict Observation Schedule

The job is both demanding and tedious. Once learned, the observational method is not difficult. Some people will find it boring and seek greater challenges. The ideal observer is one who can maintain his alertness and enthusiasm for the task, and who can find challenge in it on a day-to-day basis.

Age and sex present no inherent barriers. The majority of persons are trainable. There may be some for whom the task is too great, but there are just as likely to be some for whom the task is too easy. Most importantly, some persons will have such a fixed opinion about driving and traffic behavior (probably reflecting their own habits) that they will be psychologically unable to accept the concepts of traffic conflicts that must be used. Such persons should be identified before or during the training and given alternative assignments.

Persons presently employed as traffic technicians or paraprofessionals usually make good observers. Some agencies report that police officers may not be as good, because of their different outlook brought about by police training and experience.

A training program of 1 to 2 weeks is needed to adequately train observers. If experienced traffic technicians are utilized, the training program could be shortened.

For the current research, observers underwent a 5-day training program. The objectives of the training program were to have the observers learn the conflict definitions and the procedures for observing and recording conflicts and volumes. The film depicting staged traffic conflicts in a realistic setting that was developed during NCHRP Project 17-3 was valuable in teaching the operational definitions of the conflicts in a classroom setting. Lectures and discussions accompanied the film and were used to introduce the observation procedures.

The observers practiced observing and recording conflicts and volume at nearby signalized and unsignalized (two-way stop) intersections. Initially, the observers worked in large groups. Each day the group sizes became smaller to promote independent observation. The results of the observations were compared and critiqued during discussion sessions.

F. Observation Periods

A typical conflict study is conducted between 0700-1800 hr. The 16, 25-min observation periods are scheduled to get a representative sample of conflicts in the morning, midday, and afternoon periods of peak traffic flow. Figure 3 contains a detailed description of a typical day of conflict observation.

G. Evaluation

Traffic conflict research conducted by MRI has been accomplished in the two projects named in the introduction. NCHRP Project 17-3 included the proposal of various candidate TCT definitions and procedures, and the conduct of extensive comparative field tests. Over 9 weeks of field data were collected using 17 traffic conflict observers trained for this specific purpose. They obtained data at more than 24 intersections having a variety of geometric and traffic control configurations.

Analysis of the data collected led to a recommended set of traffic events that should be observed and recorded, together with procedures for analyzing these data. The traffic conflicts in the recommended set all have very high observer reliability. In other words, after undergoing a modest amount of training, most persons at the traffic technician level should be able to observe and record these events in nearly the same way.

The study developed preliminary estimates of the traffic conflict rates (which are better measures than traffic conflict numbers) for sites that have certain geometric configurations and traffic control devices. Further research was recommended to refine these estimates and extend them to a more comprehensive set of intersection parameters.

A procedures manual was developed for the use of agencies and traffic conflict observers planning to use the technique. This manual contains complete operational definitions and descriptions together with recording forms and detailed step-by-step procedures for the conduct of a traffic conflicts count. In addition, an instructor's guide was prepared for use in training persons in the traffic conflicts technique and in applying the technique. For detailed results, consult NCHRP Report 219.[1]

The results of the FHWA calibration study will include a set of accident/conflict relationships for specific intersection classes, described subsequently. A traffic engineer desiring to apply the technique would first conduct a conflict study. The results of the conflict study would be used to predict the expected number of accidents, by specific manner of collision, and the variance of the prediction. The traffic engineer would have an estimate of the level of safety of an intersection and also knowledge about the uncertainty of the estimate.

To develop relationships, traffic accident, conflict, and volume data were collected at 46 intersections in four cities in the Kansas City, Missouri, metropolitan area. At each intersection, 3 years (1979, 1980, 1981) of accident data, 4 days of conflict counts, and 1 day of turning movement volume counts were collected.

The intersections for the field studies were selected to provide a wide range of intersection types. A primary consideration, however, was that the test intersections have relatively simple signalization and geometrics. Intersections without sophisticated signal phasing and channelization tend to produce more conflicts than intersections with separated traffic movements. Also, intersections in the Kansas City metropolitan area, if of basic design, would tend to operate similarly to the same types of intersections in other parts of the United States. By emphasizing the more basic and common geometrics and signalization, the findings should be of more universal value.

The following general guidelines were considered for a possible test site: minimum pedestrian traffic, no unusual sight restrictions (horizontal, vertical, and cross-corner sight distance); no appreciable grade (downgrades produce more brake-light applications); no parking restrictions during just a portion of the workday ("no parking any time" restrictions were acceptable); no one-way streets; no extreme turning volumes; available observation locations on the study approaches; no oblique or offset intersections; no appreciable driveway traffic; no unusual traffic peaking; no unusually heavy truck or bus traffic; minimal congestion; and no freeway ramps.

In addition to the above requirements, the study sites were required to have a known accident history to be utilized in developing accident/conflict relationships. Obviously, some compromises were required.

Out of 200 candidates, the 46 study intersections were selected according to the following attributes:

    10 unsignalized, low volume
    10 unsignalized, medium volume
    14 signalized, medium volume
    12 signalized, high volume

The traffic volume ranges are as follows:

    low - 2,500-10,000 vehicles per day
    medium - 10,000-25,000 vehicles per day
    high - greater than 25,000 vehicle per day

At this period of the current research, work is being conducted in the analysis of the data. Results are not yet available for publication.

H.   Malmö Experiment

The procedures to be utilized during the Malmö experiment are described above in Part C - Data Collection Procedure. The only anticipated modification to the procedure will occur at the signalized intersection to be studied. The TCT will be applied using two observers instead of the usual four.

I.   References

1.   Glauz, W. D., and D. J. Migletz, "Application of Traffic Conflict Analysis at Intersections." National Cooperative Highway Research Program Report No. 219, Midwest Research Institute, Kansas City, Missouri, February 1980.

2.   Perkins, S. R., and J. J. Harris, "Traffic Conflict Characteristics - Accident Potential at Intersections." Highway Research Record No. 225, 1968.

3.   Hydén, C., "A Traffic Conflicts Technique for Determining Risk." University of Lund, 1977.

THE BRITISH TRAFFIC CONFLICT TECHNIQUE

C. J. Baguley
Transport and Road Research Laboratory
Crowthorne, Berkshire, RG11 6AU,
England.

1. Introduction

In the United Kingdom, the majority of traffic accidents involving personal injury occur at road junctions, and ways of reducing these accidents are continually being sought. Attempts to discover common reasons for accidents at individual junctions using accident records alone have proved difficult as these records often contain unreliable or inadequate information. It was considered that the concept of near-accidents or conflicts, which occur much more frequently than injury accidents, could potentially provide more information about the danger-ous aspects of a junction in a relatively short time. They could also provide comparative before and after data for determining the effective-ness of any countermeasure introduced.

This paper describes the method of traffic conflict data collection and analysis currently being used in the U.K. A training package for new observers and the results of various validation studies of the tech-nique are also summarised.

2. Conflict definitions and technique

The original method of recording conflicts reported by Perkins and Harris (1968) has been developed by TRRL for use in this country. Stud-ies to refine the technique, establish the variability of data obtained, and to determine its relationship with recorded injury accidents were initially carried out at junctions in rural areas, but in recent years more urban sites have been included.

The technique is based on observation and recording of occurrences of evasive manoeuvres, the manoeuvres being either braking or changing lane (swerving) in order to avoid a collision. The formal definition of a conflict, which was agreed at the first workshop on traffic conflicts (Amundsen, 1977), is 'an observable situation in which one or more road users approach each other or another object in space and time to such an extent that there is a risk of collision if their movements remain unchanged.

It was recognised that, depending on available time before a pos-sible collision and awareness of drivers, the severity of evasive manoeuvres varied considerably and thus an additional classification by severity of conflict was introduced (Spicer, 1971). A description of

NATO ASI Series, Vol. F5
International Calibration Study of Traffic Conflict Techniques
Edited by E. Asmussen
© Springer-Verlag Berlin Heidelberg 1984

the five severity grades at present used in the U.K. is given in Table 1

TABLE 1. Conflict severity classification

Conflict severity	Grade	Description
SLIGHT	1	Controlled braking or lane change to avoid collision but with ample time for manoeuvre
SERIOUS	2	Braking or lane change to avoid collision with less time for manoeuvre than for a slight conflict or requiring complex or more severe action
	3	Rapid deceleration, lane change or stopping to avoid collision resulting in a near miss situation (no time for steady controlled manoeuvre)
	4	Emergency braking or violent swerve to avoid collision resulting in a very near miss situation or minor collision
	5	Emergency action followed by collision

A conflict study involves two or more trained observers monitoring a junction usually from opposing directions along the priority road, and recording all relevant details of each conflict that occurs as well as subjective judgements of the evasive action taken. It has been established that four factors are used by experienced observers in assessing the grade of a conflict. These factors are:-

i)   The TIME before the possible collision that evasive action is commenced.

ii) The SEVERITY or rapidity of the evasive action.

iii) The TYPE of evasive action i.e. whether it involves more than one action.

iv) The PROXIMITY of the vehicles involved at the instant evasive action is terminated.

A relationship between judgements of the levels of these factors and conflict grades has also been established so that observers can quickly assess and record the level of each factor associated with a particular conflict (Table 2) and these can subsequently be applied to Table 3 to obtain conflict grade. This factor rating approach has proved most helpful in training new observers and has also resulted in a more consistent recording of conflicts and their severity grades. This may be due to the fact that the method promotes careful observation of several aspects of the evasive action involved in a conflict.

TABLE 2. Four-factor level ratings

Factor	Level
1 TIME before possible collision when evasive action commences	i) Long time ii) Moderate time iii) Short time
2 SEVERITY of the evasive action	i) Light braking and/or swerving ii) Medium braking and/or swerving iii) Heavy braking and/or swerving iv) Emergency braking and/or swerving
3 TYPE Whether evasive action comprises one or more types	i) Simple - either braking or swerving alone ii) Complex - both braking and swerving
4 PROXIMITY Distance between conflicting vehicles when evasive action terminated	i) More than 2 car lengths ii) Between 1 and 2 car lengths iii) One car length or less iv) Minor collision v) Major collision

To summarise, the British technique is one involving on-site observers subjectively assessing traffic conflicts. Data are collected on those situations where evasive action is taken by one or more drivers to avoid collision and these conflicts are graded according to their severity. Assessment of the levels of four factors involved in each conflict are used to determine conflict grade.

3. Data collection and recording

When a conflict situation arises the above four factors of each conflict are recorded on a standard sheet, an example of which is shown in Figure 1. In order to aid identification of hazardous features, observers also record as many details of events leading up to the conflict as possible including an accurate time of occurrence, vehicle manoeuvres involved, and any unusual features of the conflict. It is, however, essential that the time taken writing these details is kept to a minimum so that other conflicts are not missed. Junction diagrams similar to those shown in Figure 1 are used on site for observers to note down quickly the letters corresponding to vehicle manoeuvres involved. This also aids subsequent tabulation of conflict types.

Experience has shown that trained observers can deal efficiently with observation areas which have total inflows (for all manoeuvres) of up to about 1000 vehicles/hour. If more than one observer is required on a single arm of a junction then particular conflicting manoeuvres

TABLE 3. Conflict grades by factor ratings

TIME	Long		Moderate				Short			
SEVERITY	Light	Medium	Light		Medium	Heavy	Medium		Heavy	Emergency
TYPE	Simple/complex	Simple/complex	Simple	Complex	Simple/complex	Simple/complex	Simple	Complex	Simple/complex	Simple/complex
> 2 car lengths	1	1	1	1	1	2	1	1	2	3
1 to 2 car lengths	1	2	1	2	2	3	2	3	3	3
1 car length or less	2	3	2	3	3	3	3	3	4	4
Minor collision	3	4	3	3	4	4	4	4	4	4
Major collision	5	5	5	5	5	5	5	5	5	5

PROXIMITY

are specified for each observer in order to avoid double counting.

Although much higher correlation coefficients have been found between serious conflicts (grade 2 and above) and personal injury accidents (Spicer, 1971, 1972, 1973; TRRL, 1980) than between slight conflicts and accidents, details of slight conflicts are still recorded. This is because serious conflicts can be relatively infrequent events and it is considered that recording the more frequent slight conflicts helps to maintain the concentration of observers so that they are better able to distinguish, and less likely to miss, the more important serious conflicts.

The only exception to the above procedure is for the case of rear-end conflicts involving a priority road vehicle turning off (to right or left) and following vehicle(s). The driver of the following vehicle usually brakes because there is insufficient road width to pass the decelerating turning vehicle. However, the following driver normally has some warning of the intention of the driver ahead by means of indicators or brake lights and any accidents ensuing from such a conflict are usually minor involving damage only. Other researchers (Cooper, 1977; Glauz and Migletz, 1980; Lightburn et al, 1981) have shown that this particular type of conflict does not correlate well with reported accidents. Also, the levels of the factors 'time' and 'proximity' given in Table 3 are inappropriate for this combination of manoeuvres. For these reasons, observers have to apply special judgement in assessing rear-end conflicts; they are only recorded where there existed a very serious risk of collision and slight rear-end conflicts which are often very frequent are ignored.

Traffic flow counts of the various possible manocuvres at a study site are also made by means of automatic counters or additional observation methods.

4. Data Analysis

Conflict data collected by TRRL has been used in refining the technique and in various validation and investigative studies which are summarised later. Application of the technique has also been carried out by many highway authorities in the U.K. in helping to identify the problems which are most likely to lead to injury accidents at individual junctions.

One of the most useful forms of presentation of the results of conflict studies is by classifying the number of observed serious conflicts by the manoeuvres and vehicle flows involved in diagrammatic form, such as the example shown in Figure 2. The problem areas can then be studied in

detail by reference back to the original conflict record sheets to find
the ways in which conflicts occurred in order to determine patterns of
involvement.

It is frequently found that only one manoeuvre type is responsible
for generating the majority of serious conflicts at a site and, if this
agrees with available accident details, then obviously any countermeasure should be aimed at reducing this particular conflict rate. The
experience of various local authorities has shown that normally, a
suitable countermeasure (e.g. introduction of a mini-roundabout) can
completely eliminate this type of conflict.

Conflicts also provide a means of quickly evaluating the effectiveness of a countermeasure without the need to wait for long periods
in order to collect sufficient accident data. Ideally, conflict studies
should always be made before and after any junction modification is made
as they can identify any undesirable secondary effects the measure may
have had on manoeuvres other than the one it has been designed to improv
In many cases, further minor modification (e.g. an alteration in road
markings or geometry to increase deviation of path of traffic) can alle-
viate this secondary adverse effect.

5. Observer Training

i) Background

In order to obtain a reliable mean conflict count it is, of course
essential to minimise any variability in the assessments made by observers. Studies by Older and Spicer (1976) and Older and Shippey (1977)
showed that good agreement could be attained between trained observers
for both slight and serious grades of conflicts.

As already mentioned, the four-factor method was developed by TRRL
to aid training of new observers and, in response to demand from local
highway authorities, a contract was given to Nottingham University to
produce a training package using this method. Those local authorities
which already carried out conflict studies tended to use experienced
traffic engineers as observers and none had used part-time staff for
this type of observation work. It was considered that conflict studies
could be carried out much more cheaply and would be more widely used if
an easy and effective training package was available such that part-tim
workers could be trained to carry out conflict data collection reliably
Although time-lapse film has been used extensively for development of th
conflict technique by TRRL, on-site observers are again cheaper overall
and results are obtained much more quickly than with film methods.

A pilot study employing a group of students and housewives who
were asked to identify and grade conflicts from film (Lightburn and
Howarth, 1980) showed that subjects could produce reliable results which
could be improved by increasing the length of training period. By the
third day of this study (a total of 9 hours training) the correlation
coefficient between subjects was 0.68 and the mean correlation coefficient for within-subject relability was 0.75. However, this latter
figure showed considerable variance, and poor quality subjects were
shown to greatly influence the reliability of results. It was therefore
recommended that trainers should be discriminating in their final
selection of observers to be used for field studies.

Seven local authorities, who had had some experience of conflict
studies and were enthusiastic about future use, were consulted for their
opinions about the format of the proposed manual and these were taken
into account during its production.

ii)   The training package

The training package comprises a manual and 16 mm time-lapse film
containing examples of conflicts taken from real-life situations. The
first part of the manual is concerned with the rationale for conflict
studies and advice on study design. The need to use conflict data as a
supplement and not a replacement for accident investigations is stressed
and limitations of the technique are outlined. These include the subjectivity of data collected and the fact that conflicts cannot be expected
to identify all accident types (e.g. single vehicle accidents) or to
reflect accidents occurring at times of day outside the periods of conflicts observation. Conflict studies are often most useful at low traffic volume, low accident rate sites.

An introductory training booklet, included in the manual, is given
to all new trainees and explains the concept of a conflict and describes
all types of conflict which can be expected. Written answers to questions set in this booklet are completed by trainees and any problems
discussed before proceeding to the next stage of training.

The remaining section of the manual is used in conjunction with the
training film. Each of the four-factors used in assessing conflict
grade is introduced with appropriate film examples and exercises for
trainees to gain experience of identifying conflicts using the standard
recording sheet.

Finally, trainees are taken to one or two sites with relatively
high rates of conflict occurrence where further practise of conflict

recording is carried out with an experienced observer. Satisfactory
levels of performance must be attained by all trainees and if there is
any doubt as to the suitability of a potential observer then such a
person should not be used for field studies.

Even after completing the training package further reliability
checks should occasionally be made on observers by experienced observer
either on-site or from film.

Nottingham University have recently begun an evaluation project
to test the effectiveness of the package by making it available to abou
5 local authorities in both rural and industrial areas. Its use during
training sessions and subsequent field studies will be monitored closel

6. Observation periods

As road accidents can occur at any time of day, conflicts should,
perhaps, ideally be observed for the full 24 hours in order to reflect
all normally occurring traffic and lighting conditions. However, this
is obviously impractical and a good compromise, on which validation
studies have been based, is an observation period of 10 hours per day
from 08.00 to 18.00 hours. This period spans most of the morning and
evening rush hours, any minor peaks which normally occur around the
lunch period, and also relatively low flow conditions.

It has been found that as long as observers can be accommodated in
relative comfort, preferably in a vehicle parked in a suitable viewing
position off the carriageway, then satisfactory concentration can be
maintained for 5-hour periods on site per day.

The aim of site observations is, of course, to obtain reliable
mean daily conflict counts and, although day-to-day variation does
appear to depend on type of site, in most cases three days of observation is the normal period used. Spicer et al (1980) and Hauer (1978)
have shown that the increase in accuracy per additional day diminishes
rapidly and there is not much to be gained by counting for longer than
three days.

7. Summary of evaluation research

Early pilot studies by Spicer (1971, 1972) at two relatively high
accident rate sites showed good rank correlations between serious conflicts and 3-year accident records classified both by time of day and
manoeuvres involved. However, no statistically significant correlation
between slight conflicts and accidents was found. Evaluation studies
were extended to six (Spicer, 1973) and eventually fourteen sites

(TRRL, 1980) having 3-year injury accident histories which ranged from 2 to 24 accidents. The overall correlation coefficient between serious conflicts and accidents was again high (r = 0.87) and the results are shown plotted in Figure 3. At sites where there were sufficient numbers of accidents, the types of conflicting manoeuvres were ranked in order of accident and conflict rates for each site. Spearman rank correlation coefficients ranged from 0.88 to 0.97, statistically significant at the 1 per cent level, demonstrating that serious conflicts are a good indicator of the vehicle manoeuvres most frequently resulting in accidents. Again, no relation between either slight conflicts and accidents or traffic flow and accidents was found.

Owing to doubts being expressed about the day-to-day variation of conflict counts, a long-term study of one particular T-junction was made involving 27 days of observations over a period of 6 months (Spicer et al, 1980). The weekday conflict counts obtained showed no consistent day of week or seasonal effects. Day-to-day variability did however exist and, for the different manoeuvre combinations, this varied with the size of mean daily count; the following empirical relation was given:-

$$\sigma_x^2 = \bar{x}^{1.2}$$

where σ_x^2 is the variance of daily counts and $\bar{x}$ the mean daily count. Results at this site also showed that daily counts of slight conflicts were closely related to flow levels (r = 0.91). The relationship of serious conflicts with flow, however, was much less strong (r = 0.48).

Study of the variation in daily conflict counts is being extended to include 17 new sites each observed for 6 days as described at the last ICTCT meeting (Baguley, 1982). Other influencing factors such as type of site, approach speed of priority road vehicles, vehicle type, as well as accidents and flows are also being investigated at these sites.

Recently completed work by Swain and Howarth (1983) of Nottingham University has provided further evidence of the relationship of serious conflicts with accidents. Their results indicated that the relationship varied with type of site but comparing data from the most common type of junction used in the study (i.e. 8 unsignalised T-junctions), a Spearman rank correlation coefficient of 0.87 was obtained between serious conflicts per vehicle and accidents per vehicle. Total inflow was found to have non significant correlations with accidents and serious conflicts.

Various researchers (e.g. Guttinger, 1980; Malaterre and Muhlrad, 1980) have suggested that local residents' or engineers' subjective opinions of the risk at junctions can be as good a predictor of accidents as conflict counts. As rapid visual assessment would obviously provide a much cheaper way of identifying hazards, Swain and Howarth, (1983) conducted a study to test two methods of this type. The first method involved showing subjects photographs and maps of junctions from which they had to make assessments of level of risk. In the second method subjects were driven around the sites to observe each junction from different positions and also experience the level of traffic density before making their assessment. Three groups of 10 subjects (comprising ordinary drivers, driving instructors and traffic engineers) and the same junctions as the above conflict study were used in this experiment.

Data obtained from each group of drivers revealed that there was significant agreement between subjects in their individual ratings of sites both from photographs and roadside observations. This suggests that subjects were applying the same sorts of criteria when making their assessment. The junctions were ranked according to the mean subjective ratings of each group of drivers and these were compared with objective rankings based on accidents per vehicle for the first method and total accidents for the second. There was no statistically significant correlation between subjective and objective assessments of risk for any category of driver in either of the two methods. It was concluded that rapid subjective assessments such as the methods used in this study are unsuccessful in identifying hazardous road junctions and do not, therefore, provide an acceptable alternative to conflict studies.

8. Summary and Conclusions

The British conflict technique is one using on-site observers making subjective assessments of vehicle conflicts. These are based on evasive manoeuvres made by one or more drivers to avoid collision and are graded according to the severity of these events.

Serious conflicts have been shown to correlate well with recorded personal injury accidents and the technique has proved most useful to local highway authorities in identifying accident generating problems at junctions. Conflict studies can also be used to provide a relatively quick evaluation of applied countermeasures and an indication of any new problems. A training package is at present being evaluated which is to be used by local authorities to train part-time observers, thus enabling costs of studies to be kept low and promoting wider use of the technique

Further research is being carried out by TRRL to investigate the variability of conflict counts and establish general relationships of conflicts with other measurable traffic parameters. It is hoped that future research in this field will eventually enable reliable safety criteria to be specified for use by traffic engineers in the design of road junctions.

9. References

AMUNDSEN, F A, 1977. Proceedings, First Workshop on Traffic Conflicts. Institute of Transport Economics, Oslo.

BAGULEY, C J, 1982. The British traffic conflict technique: state of the art report. Proceedings, Third International Workshop on Traffic Conflicts Techniques. Institute for Road Safety Research, SWOV, Leidschendam, The Netherlands.

COOPER, P, 1977. Report on Traffic Conflicts Research in Canada. Proceedings, First Workshop on Traffic Conflicts. Institute of Transport Economics, Oslo.

GLAUZ, W D and MIGLETZ, D J, 1980. Application of traffic conflict analysis at intersections. National Cooperative Highway Research Programme Report No 219. Midwest Research Institute, Kansas City.

GUTTINGER, V A, 1980. The validation of a conflict observation technique for child pedestrians in residential areas. Proceedings, Second International Traffic Conflicts Technique Workshop, TRRL Report SR557. Transport and Road Research Laboratory, Crowthorne.

HAUER, E, 1978. Traffic conflict surveys: some study design considerations, TRRL Report SR352. Transport and Road Research Laboratory, Crowthorne.

LIGHTBURN, A and HOWARTH, C I, 1980. A study of observer variability and reliability in the detection and grading of traffic conflicts. Proceedings, Second International Traffic Conflicts Technique Workshop, TRRL Report SR557. Transport and Road Research Laboratory, Crowthorne.

MALATERRE, G and MUHLRAD, N, 1980. Conflicts and accidents as tools for safety diagnosis. Proceedings, Second International Traffic Conflicts Technique Workshop, TRRL Report SR557. Transport and Road Research Laboratory, Crowthorne.

OLDER, S J and SPICER, B R, 1976. Traffic conflicts - a development in accident research. Human Factors, 18(4), 335-350.

OLDER, S J and SHIPPEY, J, 1977. Traffic conflict studies in the United Kingdom. Proceedings, First Workshop on Traffic Conflicts. Institute of Transport Economics, Oslo.

PERKINS, S R and HARRIS, J I, 1967. General Motors Corporation Research Laboratories Research Publication. GMR-718, Warren, Michigan, also Highway Research Report No 225, HRB Washington DC.

SPICER, B R, 1971. A pilot study of traffic conflicts at a rural dual carriageway intersection. RRL Report LR410. Road Research Laboratory, Crowthorne.

SPICER, B R, 1972. A traffic conflict study at an intersection on the Andoversford by-pass. TRRL Report LR520. Transport and Road Research Laboratory, Crowthorne.

SPICER, B R, 1973. A study of traffic conflicts at six intersections. TRRL Report LR551. Transport and Road Research Laboratory, Crowthorne.

SPICER, B R, WHEELER, A H and OLDER, S J, 1980. Variation in vehicle conflicts at a T-junction and comparison with recorded collisions. TRRL Report LR545. Transport and Road Research Laboratory, Crowthorne.

SWAIN, J S and HOWARTH, C I, 1983. A comparison of subjective and objective assessments of risk at a variety of road junctions. Dept. of Psychology, University of Nottingham (To be published).

TRANSPORT AND ROAD RESEARCH LABORATORY, 1980. Traffic conflicts and accidents at road junctions. TRRL Leaflet LF918. Crowthorne.

SITE: A32/A272 West Meon   DATE: 27/5/82   WEATHER: Cloudy, road dry (a.m.) Sunny periods (p.m.)

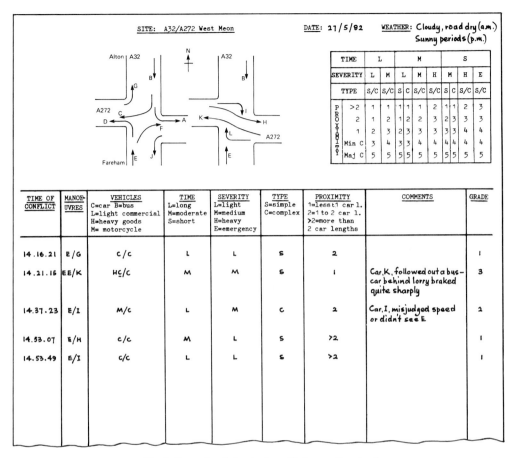

TIME	L		M			S		
SEVERITY	L	M	L	M	H	M	H	E
TYPE	S/C	S/C	S	C	S/C	S/C	S	C	S/C	S/C

PROXIMITY										
>2	1	1	1	1	1	2	1	1	2	3
2	1	2	1	2	2	3	2	3	3	3
1	2	3	2	3	3	3	3	3	4	4
Min C	3	4	3	3	4	4	4	4	4	4
Maj C	5	5	5	5	5	5	5	5	5	5

TIME OF CONFLICT	MANOE-UVRES	VEHICLES C=car B=bus L=light commercial H=heavy goods M= motorcycle	TIME L=long M=moderate S=short	SEVERITY L=light M=medium H=heavy E=emergency	TYPE S=simple C=complex	PROXIMITY 1=less t1 car l. 2=1 to 2 car l. >2=more than 2 car lengths	COMMENTS	GRADE
14.16.21	E/G	c/c	L	L	s	2		1
14.21.16	EE/K	Hc/c	M	M	s	1	Car,K, followed out a bus-car behind lorry braked quite sharply	3
14.37.23	E/I	M/c	L	M	c	2	Car,I, misjudged speed or didn't see E	2
14.53.07	E/H	c/c	M	L	s	>2		1
14.53.49	E/I	c/c	L	L	s	>2		1

Fig. 1 Example of standard conflict recording sheet

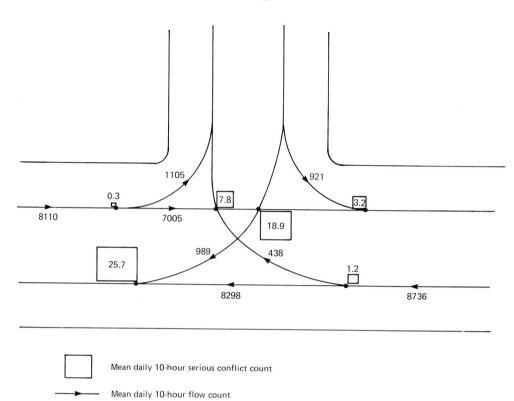

Fig. 2 Example of diagrammatic representation of conflict counts and vehicle manoeuvres flows

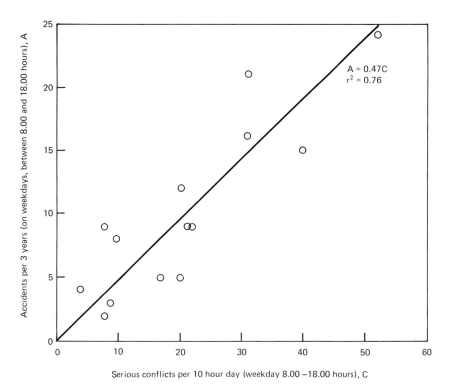

Fig. 3   Reported personal injury accidents and observed conflicts
at fourteen intersections

EXPERIENCE WITH TRAFFIC CONFLICTS IN CANADA
WITH EMPHASIS ON "POST ENCROACHMENT TIME" TECHNIQUES

P.J. Cooper
Insurance Corporation of British Columbia
Vancouver, B.C., Canada

---

1. Introduction

Following initial investigation of the G.M. (brakelight) conflict
recording technique during 1972/73 in various Canadian cities,
it was concluded that this procedure - while relatively easy to
apply had some very fundamental conceptual drawbacks as an indicator of expected accident rate.

Accordingly, a study of actual film-recorded accident sequences
was undertaken in an attempt to define the relevant, measurable
parameters involved and to see whether or not a more theoretically
acceptable link between accidents and "conflicts" could be established. The conflict definition arising from this work was referred
to as "Post Encroachment Time" (PET).

Initial evaluation of the PET technique was undertaken during 1978/
79 using only urban, signalized intersections. Later, during 1980/
81, additional work was done during which non-signalized intersections were also evaluated. The performance of PET was assessed
against the brake light technique and simple volume-related exposure models.

2. Definitions and conflict types

A traffic conflict is an event involving two or more vehicles in
which the unusual action of one vehicle, such as a change in speed
or direction, places the other vehicle in danger of a collision
unless an evasive maneuver is taken. Traffic conflicts do not
include vehicle actions that result from obeying a traffic control
device. For purposes of this project, vehicles are to include
automobiles, motorcycles and bicycles.

For a traffic conflict to occur, an actual impending vehicle collision is not necessary. A vehicle action or maneuver that threatens another vehicle with the possibility of a collision is sufficient to be considered a conflict. An intersection conflict can
be described as an event involving the following stages:

NATO ASI Series, Vol. F5
International Calibration Study of Traffic Conflict Techniques
Edited by E. Asmussen
© Springer-Verlag Berlin Heidelberg 1984

DEFINITIONS

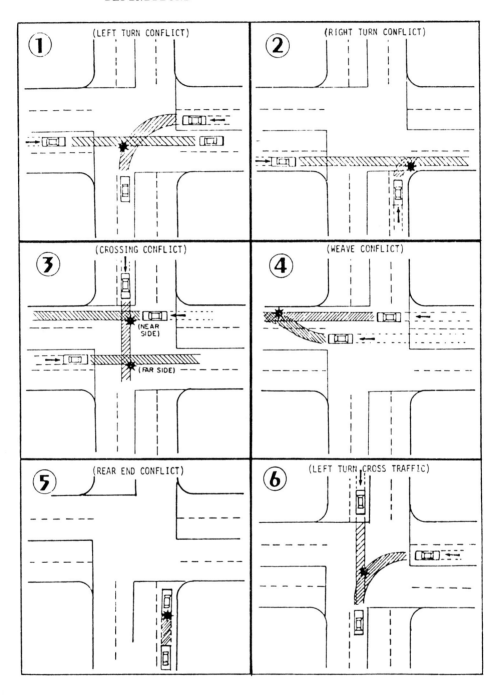

Note: Arrows indicate direction
of vehicle movement.

Conflict Types

Stage 1.   One vehicle makes an unusual or unexpected maneuver.

Stage 2.   A second (conflicted) vehicle is placed in danger of
           a collision.

Stage 3.   The second vehicle reacts by braking or swerving.

Stage 4.   The second vehicle then continues to proceed through
           the intersection.

The different types of vehicle conflicts can be defined in terms
of the above stages.

Only vehicle conflicts occurring on the major approaches of an
intersection are considered. The major approaches are generally
those having the higher traffic volumes.

Vehicle conflicts are defined in terms of the post encroachment
time (PET) concept. The basic definition of a post encroachment
time (PET) traffic conflict can be expressed as the time difference
between the moment an "offending" vehicle passes out of the area
of potential collision and the moment of arrival at the potential
collision point by the "conflicted" vehicle possessing the right-
of-way.

Left-turn and crossing situations are considered an angle type
conflict, as opposed to parallel type conflicts such as rear-end,
right-turn and weave.

Definitions of PET for angle type conflicts are relatively straight-
forward. The parallel type definitions are more difficult to establish in such a manner that totally objective observations can be
made. To this end, three constituent elements in the general PET
definitions must be considered:

1)   the potential collision point;
2)   the moment of encroachment termination, and,
3)   the location of the collision point.

Detailed PET definitions for the various types of traffic conflicts
to be observed in this project are presented below.

Left Turn, Cross-Traffic Conflict (LTCT)

A left-turn, cross-traffic conflict (LTCT) is defined as the time
difference between the moment a left-turning vehicle from the minor

approach fully enters the lane occupied by a major approach through vehicle, and the moment of arrival by the through vehicle at the potential collision point.

Opposing Left-Turn Conflict (OLT)

An opposing left-turn traffic conflict (OLT) is defined as the time difference between the moment a left-turning vehicle leaves the lane occupied by an oncoming major approach through vehicle, and the moment of arrival by the through vehicle at the potential collision point.

Right-Turn Conflict

A right-turn traffic conflict is defined as the time difference between the moment a right-turning vehicle fully enters the lane occupied by a major approach through vehicle, and the moment of arrival by the through vehicle at the potential collision point.

Crossing Conflict

A crossing conflict is defined as the time difference between the moment a crossing-vehicle leaves the lane occupied by a major approach through vehicle and the moment of arrival by the through vehicle at the potential collision point.

Weave Conflicts

A weave conflict is defined as the time difference between the moment a lane-changing vehicle fully enters the lane occupied by a major approach through vehicle travelling in the same direction, and the moment of arrival by the through vehicle at the potential collision point.

Rear-End Conflict

A rear-end traffic conflict is defined as the time difference between the moment a leading through vehicle on the major approach clears the potential collision point (the point where it either releases its brakes or comes to a stop), and the moment of arrival by the following vehicle at the potential collision point.

For a rear-end conflict to occur, it is not necessary that the leading through vehicle on the major approach stop before exiting the through lane. The PET definition would still apply as the time difference between the moment the leading vehicle exits the through lane, and the moment of arrival by the following vehicle at the potential collision point.

If a vehicle is stopped behind a left-turning through vehicle on the major approach which is waiting to complete the turn, this situation is not considered a rear-end conflict.

"Critical encroachment" is a term applied to the more critical
PET's of shorter duration which previous study has indicated have
a possible greater influence on the occurrence of collisions.

A critical encroachment event is defined as a "near miss", or other
critical situation, in which a vehicle is required to come to an
abrupt stop with abnormal deceleration behind another vehicle, regardless of the action of the vehicle ahead. Since PET measurement
does not apply in such a situation only the occurrence can be recorded.

Should a following vehicle decelerate and pull into an adjacent lane
to avoid the vehicle ahead, that situation is considered a potential
weave conflict in the adjacent lane, and is recorded by the observer
responsible for weave conflicts.

3. Observation techniques

The data team for each intersection is divided into two groups (Group
A and Group B) with the observers in each group being responsible for
particular types of traffic conflicts, or tasks. Each observer in
the group is responsible for the same task at every intersection
studied within a given project. The retaining of a task by each observer is intended to ensure familiarity and consistency with the
particular type(s) of conflict measurement in the task.

Each data team consists of 3 or 4 observers divided into a Group A
and a Group B as illustrated on the accompanying diagrams and individual observer tasks are assigned within each group.

In addition to the traffic conflicts survey, both traffic volume
counts and turning movement counts are conducted concurrently with
each PET survey period, or periods, for each intersection. The
traffic counts would normally be conducted by staff other than the
traffic conflicts observers.

The period of conflict counting is 10½ hours - usually between 0730
and 1800 hrs. This would normally require 2 working days to complete
or perhaps longer for a complex 4-way intersection.

Grouping and Positioning of Data Team about Four-Way Intersections.

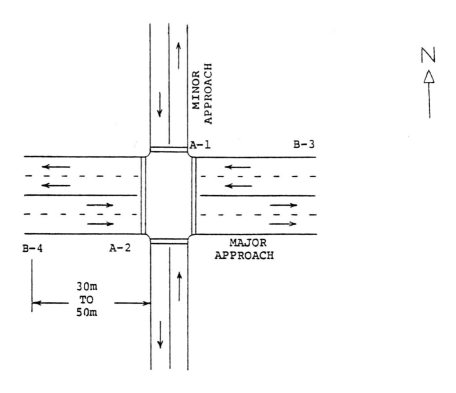

GROUP	OBSERVER	TASK
A	A-1	Measures opposing left-turn (OLT) right-turn and crossing conflicts westbound approach.
	A-2	Measures opposing left-tun (OLT) right-turn and crossing conflicts eastbound approach.
B	B-3	Measures rear-end, weave and left-turn, cross-traffic (LTCT) conflicts on westbound approach.
	B-4	Measures rear-end, weave and left-turn, cross-traffic (LTCT) conflicts on eastbound approach.

Grouping and Positioning of Data Team about Two-Way Intersections.

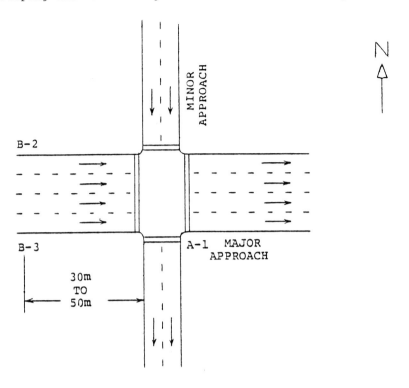

GROUP	OBSERVER	TASK
A	A-1	Measures right-turn[1] and crossing conflicts on eastbound approach.
	B-2	Measures rear-end and left-turn, cross traffic (LTCT)[1] conflicts on eastbound approach.
B	B-3	Measures weave and left-turn, cross-traffic (LTCT)[1] conflicts on eastbound approach.

NOTE 1. Task dependent upon the relative directions of traffic on the approaches.

4. Data collection format

A Traffic Conflicts Survey Form is used by the data team observer
to record each PET event. In addition to the form heading information, the types of conflict to be observed are entered on the form
in advance of the survey period. The position of the observer is
indicated on the intersection diagram in the upper right corner of
the form.

The start time of day is recorded in 15-minute increments during
the survey period, using the 24 hr. clock.

Each PET event is recorded as an event mark under the appropriate
0.5 second increment column of the "PET Range", leaving enough
space to enter the number of the totalled events within each time-
block for the successive 15-minute periods of the survey. For example, only events having PET values within the range of 3-3.4 seconds,
inclusive, are to be recorded in the column headed 3.0.

Under each "Conflict Type" heading there is a column denoted "CE",
which signifies critical encroachment. Each critical encroachment
event is recorded in the appropriate time-block in a manner similar
to the recording of PET events.

The "COMMENTS" area of the form is provided for the observer to note
any unusual events which may have interfered with normal traffic operations at the intersection. Such events might include stalled vehicles and vehicle collisions. The time of day of the event should
also be noted. The weather, visibility and road surface conditions,
including any changes in these conditions during the survey period
should be noted in the "COMMENTS" area.

The measurement of traffic conflicts is based on recording the post
encroachment time (PET) for each conflict event occurring at an intersection during a given period of the day. The following information is required to record each conflict event:

1.   the conflict type;
2.   the time period; and
3.   the observed PET in seconds.

The conflict type can be determined according to the PET definitions
for traffic conflicts contained in Part 11 of the procedures manual.
The time period is the "count start time" for each 15-minute increment during the survey. The observed PET is that time difference
measured with a stopwatch (to the nearest 0.1 second) between the
moment an "offending" vehicle passes out of the area of potential
collision, and the moment of arrival at the potential collision point

Location No. _____

Date _____

Intersection _____

Approach _____

Observer No. _____

Observer Name _____

CONFLICT TYPE

COUNT START TIME FOR 15 MIN. INCREMENTS (MILITARY)	CE	PET RANGE (SEC. ≥)							CE	PET RANGE (SEC. ≥)							CE	PET RANGE (SEC. ≥)					
		0.5	1.0	1.5	2.0	2.5	3.0			0.5	1.0	1.5	2.0	2.5	3.0			0.5	1.0	1.5	2.0	2.5	3.0

COMMENTS

PAGE _____ OF _____

TRAFFIC CONFLICTS SURVEY FORM

by the vehicle possessing the right-of-way.

Therefore, to measure and record the PET for traffic conflict event, it is necessary for the observer to:

1. determine the type of conflict;
2. time the PET for the conflict;
3. record the stopwatch reading as an event mark in the appropriate 0.5 second time-block area of the survey form; and
4. reset the stopwatch for the next conflict event.

Due to the amount of time required to record a PET measurement using the above procedure, it may be difficult to time two or more conflict events which occur very close together, or simultaneously. Since each observer is equipped with only one stopwatch, it may be necessary for the observer to estimate the PET's for events which are observed, but which cannot be measured.

Also, in peak traffic periods, the observers may have to memorize a number of consecutive PET values before recording them on the survey form, so as to minimize the number of missed conflict events.

To record a critical encroachment, it is necessary only to enter an event mark in the appropriate area on the survey form, under the designated column for the applicable type of conflict and opposite the appropriate time period. Sufficient space should be left in each blocked area to enter the number of the totalled events for the successive 15-minute periods of the survey.

5. Data treatment and evaluation

Since the Traffic Conflicts Technique has not yet become an operational tool in Canada, the treatment of conflict data has not been standardized into a routine procedure or format. The following two sections will therefore describe the analysis and results arising out of the two most recent Canadian studies - one conducted during 1978/79 and the latest during 1980/81.

5.1 1978/79 Study

Seven signalized intersections in Hamilton, Ontario were studied. The accident variable employed was the mean yearly collision rate based on six years of record and these data were considered in two forms: 24 hr. totals and only those accidents occuring between the hours of 0730 and 1800. In addition, accidents which occured in bad weather conditions were eliminated from the data set.

Traffic conflicts were observed during three time periods for a total of 10½ hours: 0730 - 1000, 1000 - 1530 and 1530 - 1800. Counts were conducted only under good weather conditions. The major conflict variable was the number of PET's less than 4 secs. recorded by conflict type and intersection approach. There were eight PET categories from 0 to 4 secs. in 0.5 sec. ranges. Some 8000 such PET conflicts were recorded and, in addition, about 10,000 brakelight observations were made.

Conflicts and collisions were related using standard linear regression techniques, with the general results discussed below.

Unique COLLISION-PET relationships during different time periods of the day could not be identified from the results of the analysis. Left turn and rear end conflicts did not exhibit strong relationships at all, and weave conflicts exhibited only marginally higher correlations but when the observation day was treated as one time period, stronger COLLISION-PET relationships were observed. In general, utilizing all 24-hour collisions in the analysis resulted in stronger correlations than those obtained when collisions were restricted to the shorter observation day (0730 - 1800).

When the collision variable was restricted to only dry weather collisions, left turn and rear end conflicts showed significantly higher correlations than cases where total 24-hour collisions were employed.

'Optimum' PET upper limit values for each case were established. These were the upper limit values that produced the strongest correlations for each particular analysis attempted, and could be intepreted as identifying the PET definition that resulted in the strongest COLLISION-PET relationship. With a few exceptions, it was noted that PET $\leq$ 2.0-2.5 seconds seemed to be a reasonably consistent choice. In five of six cases for left turns the optimum PET upper limit value was always in the 1.5-2.5 second range. This could be interpreted to indicate that the more critical PET's (shorter duration) have a greater influence on the occurence of collisions.

The slope of the COLLISION - PET relationships were always positive for rear end and weave conflicts, but were negative for left turn conflicts. Since the intersections were signalized, left-turn manoeuvers were the only crossing type conflicts recorded. This clearly suggest that the occurence of left turn (or crossing) collisions may decrease with increases in the number of left turn (or crossing) conflicts taking place. Although such an event is initially counterintuitive, one could perceive that increasing frequency would result in greater driver caution with the observed result of decreased frequency of collision.

Brakelight data were also recorded and compared with the accident

history. The results of this analysis ranged from extremely low
correlations for left turn conflicts to moderately significant values
for rear end conflicts to highly significant values for weave conflicts.

Although these results are generally encouraging, it is important
to note that the data bases for the left turn and weave brakelight
conflicts were quite small in comparison to the rear end cases.
Although no explicit rationale can be presented for this phenomenom,
field experience showed that left turn and weave conflicts often
occured without brakelight application and that occurence of such
cases far exceeded instances where brakes were applied. It is also
interesting to note that consideration of dry weather collisions did
not increase the explanatory power of the relationships.

An additional analysis was attempted to aid in explaining the
COLLISION-PET relationship: the cumulative distributions of PET's
by time range for each intersection approach and conflict type was
prepared. With all things considered equal, one could intuitively
suspect that intersection approaches with a history of higher collision rates would exhibit higher numbers and/or percentages of high-
risk PET conflicts (PET's less than 1.5 seconds, say) than those
approaches with a history of lower collision rates. Despite such
optimistic intuition, the distributions for left turn conflicts did
not exhibit such trends to any great extent. The distributions for
rear end conflicts did show promising trends, but the low number of
collisions occuring on the majority of approaches made the trends
difficult to assess and therefore reduced confidence. A similar conclusion could be drawn for weave conflicts, where even fewer accidents occured and identifiable trends were difficult to distinguish.

In addition to the standard two variable linear regression analyses
described earlier, multiple regressions were performed using severe
conflicts (0-1.5 sec.) moderate conflicts (1.5-2.5 sec.) and minor
conflicts (2.5-4.0 sec.). Brakelight application conflicts had been
already categorized in this fashion. The results indicated that the
coefficient for severe conflicts was usually larger than those for
moderate and minor conflicts, suggesting a more significant role in
the relationship.

Based on the above analyses some general conclusions were reached.
First, no significant problems were encountered with the size of
field crew employed. For all practical purposes, a 100% sample was
obtained for all conflict types, at two intersection approaches using
three persons. The manual mode of collection utilizing stopwatches

and clipboards proved to be completely acceptable once field personnel had gained a few days of experience. In fact, some crew members
became so adept at recognizing PET conflicts that visual estimations
of occurence duration were astonishingly precise.

In general then, it was concluded that PET traffic conflicts can
be collected in sufficient quantity to be considered a representative
sample at a four-legged intersection with the expenditure of a maximum of six-man days of effort.

There is considerable indication that the TCT can be used as an indicator of the degree of hazard present at an intersection. More
particularly, it would appear that the results point toward a meaningful and consistent relationship between conflict and collision
frequency. Although examination of the results revealed obvious
inconsistencies, the project team generally concluded that the PET
traffic conflicts technique can be effectively used as a safety countermeasure tool. Until such time as more data becomes available
however, specifically identified conflict/collision ratios will be
subjected to relatively low degrees of confidence.

The results surrounding the comparison of PET conflicts vs. brakelight conflicts were also interesting. When one considers all aspects
of the two methods, including ease and objectivity of data collection, it was concluded that the PET traffic conflicts technique compared favourably to the currently employed methods utilizing primarily brakelight application.

The validity of employing PET conflicts as an accident predictor was
investigated using analysis of variance techniques. Variances between actual and expected annual number of accidents were calculated
based on a formula developed by Dr. Ezra Hauer for analyzing significance of accident reduction in Metro Toronto during 1970/74.
Variances were also calculated for the expected annual collision
rate based on the occurence of conflicts - once again using formuli
developed by Dr. Hauer.

A comparison was made between the two estimates of variance for each
of the major conflict types where sufficient data existed. For rear-
end situations, the estimated variance between actual and expected
yearly accidents based on analysis of collision history was less than
that arrived at through consideration of PET conflict counts in only
about 30% of the cases (intersection approaches) and this did not
significantly change even when six years of collision history were
considered. For left-turn and weave configurations, however, the
relative performance of accident history as an expected value pre-

dictor improved significantly over conflicts as the period of accident record increased from 2 to 6 years. With left-turns, the accident records produced smaller variances in only about 25% of the
cases when just two years were considered but this proportion rose
steadily to 75% for six years of record. Weave situations went
from 6% to 44% over the same record length range.

The general conclusion which one might draw from the above is that,
on average, the PET conflicts were at least as good predictors of
expected accident rate as historical accident records of up to 6
years duration for the case of weave and rear-end situations and
that for left-turn (crossing) incidents, conflicts were as good or
better than accident history up to the point where 3 years of record
were obtainable. The influence of years of record is of course
totally dependent upon the average number of accidents occuring per
year.

5.2 1980/81 Study

Five non-signalized intersections were studied in Ottawa, Ontario
for which six-year accident histories were compiled giving a total
of 231 collisions.

As in the 1978/79 study, PET conflicts were recorded and linear
correlations between the accident and conflict data sets were examined.

In examining the cumulative frequency distribution of all PET's
measured by intersection, it appeared that all intersections were
represented by relatively consistent distributions over the time
frames measured. There did thus not appear to be any significant
difference of PET distributions by intersections.

Basically the distributions fell into two categories. For left turn,
right turn and crossing conflicts, the number of PET's measured increased exponentially as the time frame increased. For left turn
cross traffic, rear end and weave conflicts, the number of PET's
measured accumulated rather consistently by time frame up to a value
of 2.0 seconds. It is implied from this distribution that for
these latter types of conflicts, little value would be gained by
measuring PET's where the time value exceeds 1.5 or 2.0 seconds. The
optimum PET upper limit values established by best correlation with
accidents ranged from 1.0 secs. for weave conflicts to 2.5 secs. for
rear-end and turning conflicts. The mean value was about 2.0 secs.

The results of the linear regression analyses can be summarized by
conflict type as follows:

1980/81 - Study

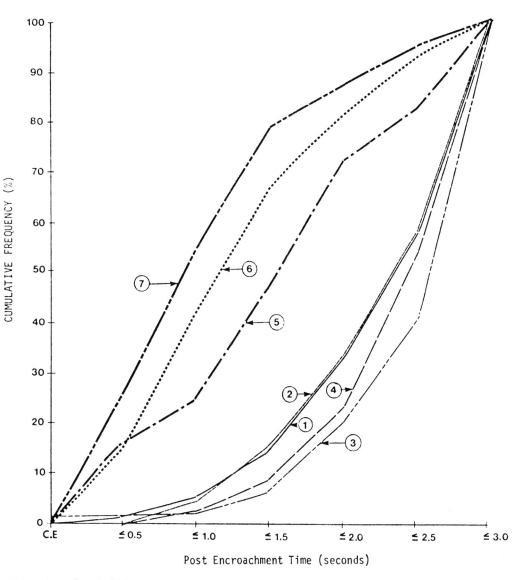

Cumulative Distribution
of Conflicts by Type.

Legend: 1 - Left Turn
         2 - Right Turn
         3 - Crossing (Near Side)
         4 - Crossing (Far Side)
         5 - Left Turn Cross Traffic
         6 - Rear Ends
         7 - Weave

Opposed Left Turn

A high degree of positive correlation was obtained for opposed left
turn conflicts, with the maximum correlation occuring when PET's
of duration less than 2.0 to 2.5 seconds were considered. The data
base for this conflict appeared to be sufficient to place reasonable
confidence in this correlation.

Right Turns

As with opposed left turns, right turns showed high degrees of positive correlation especially when PET values in the range of 2.0 to
2.5 seconds were considered. Due to the limited data base available
for analysis, these correlations should be accepted with limited
confidence.

Crossing (near side)

With one exception, all the correlation coefficients for the crossing
conflict (near side) were negative. These results are unduly influenced by the values obtained for one particular intersection. A
positive correlation coefficient was obtained when the PET upper
limit value as 1.5 seconds.

Crossing (far side)

Low levels of correlation existed for this type of conflict, except
where the PET upper limit was 1.5 seconds. Even in this case, the
correlation coefficients of 0.52 and 0.63 for total collisions and
daytime collisions respectively were not significant. As the PET
upper limit was increased, the correlations became negative. These
negative correlations resulting from inclusion of high limit PET's
may indicate a patter of "risk compensation" on the part of drivers
when the danger is readily identifiable and there is ample time for
evasive action.

Weave

Due to the limited number of accidents that related to the weave
conflict, it was considered that the values obtained did not represent a proper test of the validity of Post Encroachment Techniques
to identify safety problems related to weaving. Nevertheless, some
highly positive correlation values were obtained when total relevant
collisions were considered. This would imply an inter-relationship
between conflicts and collisions of this type.

Left Turn Cross Traffic

As with weave conflicts, it was considered that the data base for
left turn cross traffic conflicts was not sufficient to provide a

valid test of the Post Encroachment Techniques as a traffic safety
tool for this type of accident. As with weave conflicts, some highly positive correlation values were obtained again implying an inter-
relationship between conflicts and collisions of this type.

Rear End

While there were a large number of rear end conflicts measured at
each intersection, and the number of rear end collisions varied from
1 to 7, meaningful positive correlations between the two variables
were simply non-existent and therefore an interrelationship between
rear end collisions and conflicts could not be implied.

As with left-turn conflicts at signalized intersectiions in the 1978/
79 study, crossing conflicts in this investigation had strong negative correlations with associated accidents. Again, this might be
attributed to driver caution resulting from a well-perceived hazard,
but it should be noted in this case that the negative correlation
was in great part due to the position of one particular intersection
on the plot. Without this data point, the low PET correlation would
have been low-to-moderate but positive. There was no objective
reason found, however, to justify disqualification of this particular location and it did not appear anomolous in other conflict type
plots.

Rear End conflicts generally displayed the lowest degree of correlation of all conflict types considered. It was noted that the number
of rear end conflicts measured were 5 - 6 times greater than any
other conflict type and this would lead to the suspicion that a
number of precautionary events were included which had virtually no
potential to result in accidents.

Traffic volume "flow cross-product" factors were established for
all conflict situations and correlated with the accident records.
In most cases the results were significantly inferior to those obtained with the PET and this finding suggests a degree of validity for
the concept of conflicts in general and PET in particular.

Based on the data available for analysis, it was considered that the
Post Encroachment Technique offered potential for indicating traffic
safety problems at unsignalized intersections in that several conflict types produced correlations at the 5% level of significance,
particularly at the lower ranges of PET values. However, due to the
limited data base analyzed it was difficult to accept these correlations with a high degree of confidence.

1980/81 - Study

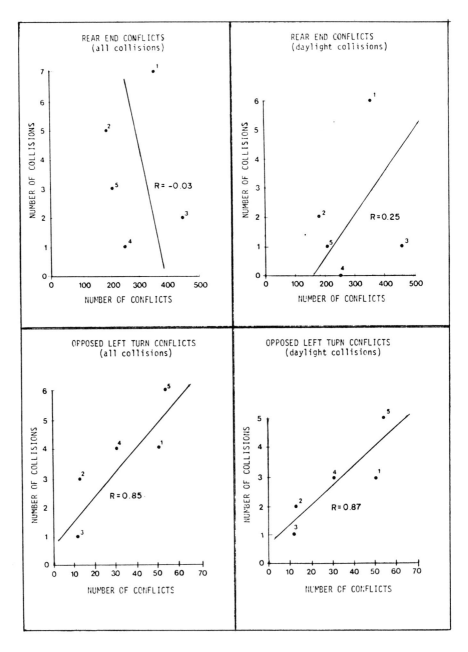

Plot of Conflict types against
Collision types.

1980/81 – Study

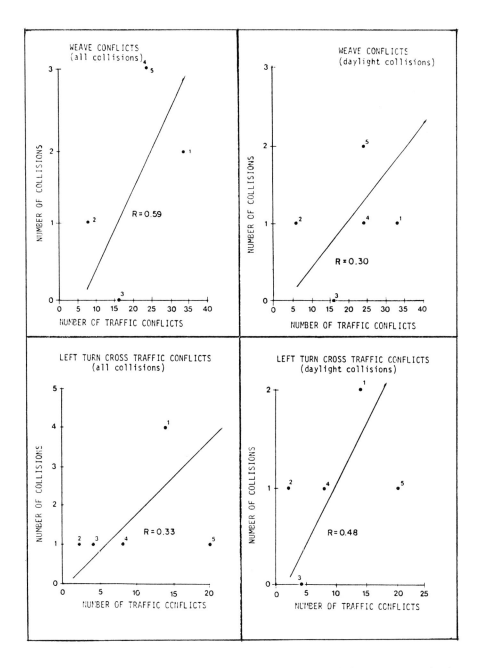

Plot of Conflict types against
Collision types.

1980/81 - Study

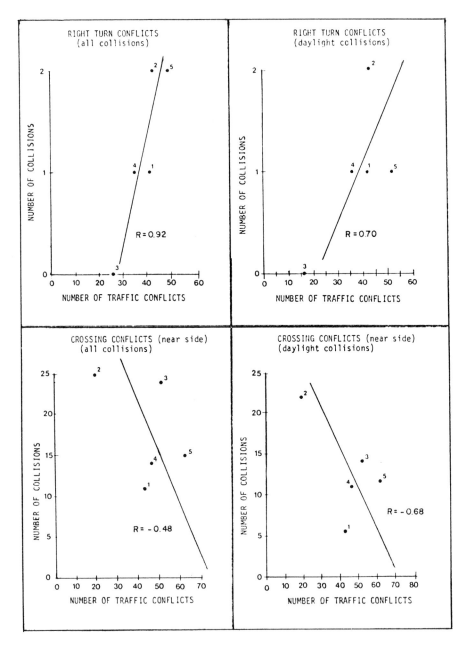

Plot of Conflict types against
Collision types.

1980/81 - Study

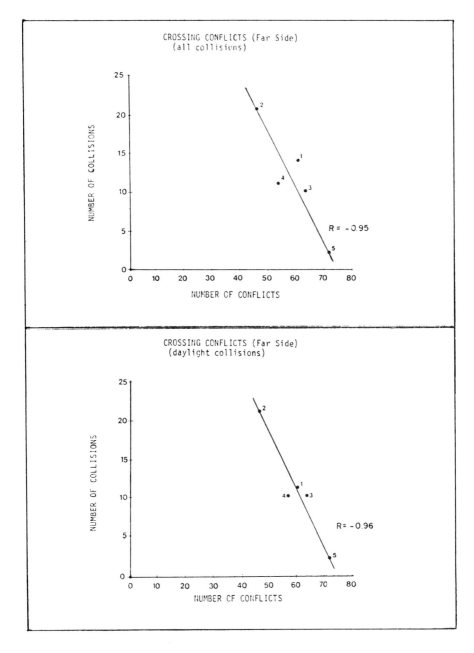

Plot of Conflict types against
Collision types.

6. Conclusions

Much of the investigative work on traffic conflicts (PET) in Canada
has suffered from small data sample sizes - especially in terms of
recorded accident frequency. The results of the latest two studies
which have been reported here can therefore at best be described
as inconclusive. Larger numbers of study locations might assist
in establishing PET validity but the major experimental design
improvement would be to utilize the technique at locations having
higher overall accident frequencies.

One of the major positive findings of the research to date has been
the establishement of PET conflicts as generally better predictors
of expected accidents than either past collision history or volume
exposure factors.

References

1. Hauer, E. and Cooper, P.J., "Effectiveness of Selective Enforce-
   ment in Reducing Accidents in Metropolitan Toronto, Transportation Research Record 643, TRB, Washington, 1977.

2. Hauer, E., "The Traffic Conflicts Technique - Fundamental Issues",
   Dept. of Civil Engineering, University of Toronto, 1976.

3. Hauer, E., "Design Considerations of Traffic Conflict Surveys",
   Transportation Research Record 667, TRB, Washington, 1978.

4. Allen, B.L. and Loutit, C.B., "Investigation of Post Encroachment Time as a Traffic Conflicts Technique", Dept of Civil
   Engineering, McMaster University, Hamilton, Ontario, 1979.

5. Damas and Smith Ltd., "Post Encroachment Time Conflict Technique -
   A Traffic Safety Tool ?", Final Report to Transport Canada, Ottawa,
   1981.

THE FINNISH TRAFFIC CONFLICT TECHNIQUE

Risto Kulmala
Road and Traffic Laboratory
Technical Research Centre of Finland
02150 Espoo 15/Finland

Introduction

The traffic conflict technique was introduced in Finnish traffic
safety research by Roads and Waterways Administration in 1972. The
technique was based on observing brakings on the major road at rural
junctions. The first conflict studies at urban junctions were made by
the Technical Research Centre of Finland in 1974. The collected conflict data was mainly used in developing a conflict simulation model.

No conflict studies were made at the Technical Research Centre until
1979, when the Road and Traffic Laboratory was commissioned to make
conflict observation in the cities of Helsinki and Lahti. After these
studies we were able to convince authorities of the usefulness of the
technique especially in short-term evaluation of safety measures.

The conflict technique used by the Technical Research Centre of Finland is basically the technique developed by Christer Hydén in Sweden
with some modifications. These modifications were made on the basis
of experiences from our first studies and they were made in order to
improve the quality of the collected observation data.

Definitions

Situations, where braking or weaving begins 1,5 sec or less before
a potential collision, are defined to be conflicts. If braking or
weaving is uncontrolled the conflict is defined serious. This definition is used in urban areas, where the speed limit is 50 km/h. The
time-to-collision value varies with the used speeds so that the value
is 3,0 sec when the speed limit is 100 km/h. When conflicts between
a cyclist and a pedestrian or between two cyclists are studied, the
time-to-collision value is 1,0 sec.

Potential conflict situations are defined to be situations, where the
participants adjust their speeds well enough before the potential
collision. All participants don't however, behave in a way required,

NATO ASI Series, Vol. F5
International Calibration Study of Traffic Conflict Techniques
Edited by E. Asmussen
© Springer-Verlag Berlin Heidelberg 1984

and the situation nearly ends up in a conflict. The definition is rather a subjective one and we are now working on a more objective definition based on time-to-collision values.

Types of road situations observed

All conflicts and potential conflicts where at least one of the participants is driving a motor vehicle or bicycle, are observed. We also register all observed traffic violations in the study location.

In almost all of our studies also other situations have been observed. The types of these situations vary as they are very specific and directly linked to the problem studied. For example when the safety of climbing lanes was studied, we registered overtakings, lane changes etc.

Observation means

We always have observers on location. In addition, video equipment is nearly always used in order to check the observations and also collect exposure and other relevant data.

Data collection and forms

Each observer marks his observations on two forms: the conflict form and the potential conflict form. These forms are shown on figures 1 - 3. On the conflict form each conflict is described by a drawing of the situation. The severity of the conflict is marked by a cross and other important information is written beside the drawing. The exact time is also written down for later checking.

Each potential conflict situation is marked in the corresponding box of the potential conflicts form depending on the junction leg and the type of the situation. The time, when the situation occurred, is written in the box.

Traffic flows are usually gathered from the video tapes later in the laboratory on a special form, which is also shown on the next pages. The traffic flow data is gathered in samples of 5 minutes, totalling 25 - 50 % of the observation period. Pedestrians, bicyclists, two-wheeled vehicles and other motor vehicles are counted separately, as shown on the form.

VTT road and traffic laboratory

date____________ time ____________

location________________________

CONFLICT OBSERVATIONS

observer____________ page________

1
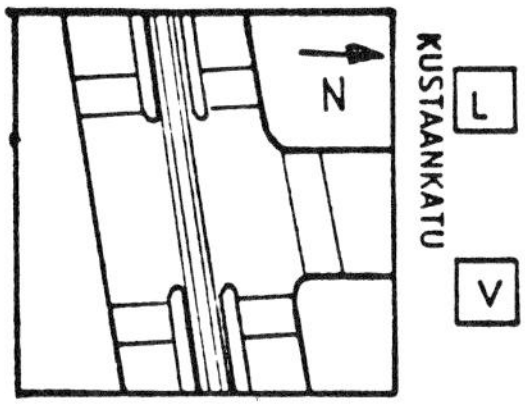

huom:

2
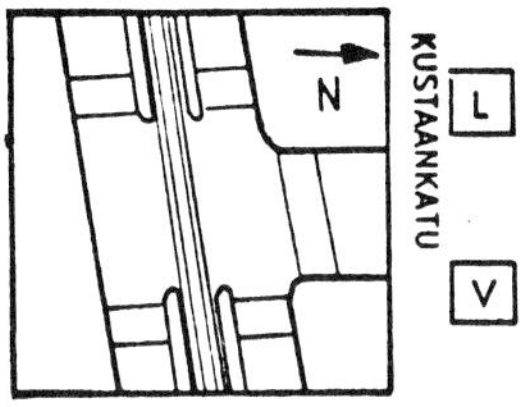

huom:

3
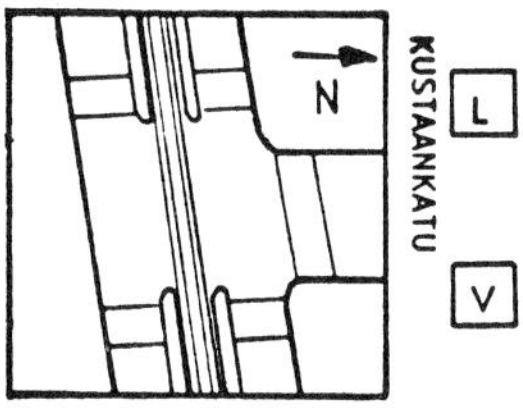

huom:

4
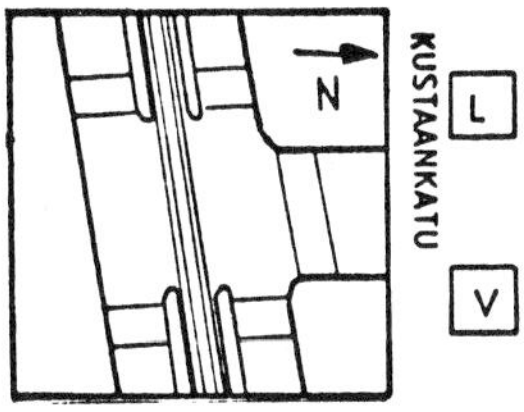

huom:

5
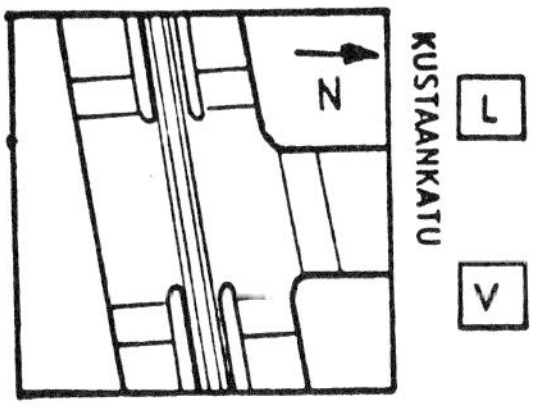

huom:

Figure 1. Conflict form.

CONFLICT STUDIES   POTENTIAL CONFLICT SITUATIONS FORM

Junction/location _____ The number of north leg _____

Date _____. _____.19___, time_____. _____ - _____. _____

Additional information _____

Traffic situation	Number of junction leg or pedestrian crossing				Total
	1	2	3	4	
Wrong driving order					
Right of-way conflict from the left					
" from the right					
" opposing left turn					
Rear-end conflict to left					
" to right					
" straight through					
Other motor vehicle situation					
Pedestrian conflict					
Traffic violation, not including the following 3					
Pedestrian crosses street outside crossing					
Pedestrian against red					
Motor vehicle against red					
Other situation					
	1	2	3	4	

Figure 2. Potential conflict situations form.

CONFLICT STUDIES   TRAFFIC FLOWS FORM

Junction/location _____   The number of north leg _____

Date_____._____.19___ , time_____._____. -_____._____

Additional information_____

Junction leg	Flow	Cars	Total	Two-wheeler	Total
1	↙				
1	↓				
1	↘				
2	↖				
2	←				
2	↙				
3	↗				
3	↑				
3	↖				
4	↗				
4	→				
4	↘				

Pedestrian crossings		Pedestrians	Total	Cyclists	Total
	1				
	2				
	3				
	4				

Figure 3. Traffic flows form.

Data treatment

When analysing conflict and potential conflict situation data we always compare it to exposure data.

Exposure is usually calculated as a square root of the product of conflicting traffic flows or in some cases the number of road users in a
traffic stream or passing through the study location during the observation period.

We calculate then conflict risks (rates) by dividing the number of conflicts by exposure. Potential conflict situation risks are calculated
in the same way. The significance of risk differences (between locations or before-after) are tested with χ^2-test.

We also study the location's accident material if it is available and
compare it with the material obtained in our observations. The accident material enables us to predict future accident numbers in before-
after studies on the basis of observed changes in conflict risks.

Based on conflict frequencies and risks and also the observers'
opinions we give recommendations concerning possible safety measures
on the study location.

Most of our studies are commissioned by Finnish towns and their
traffic planners. Because of that we try to present our results in a
way that would give the planner a clear picture of the location's
safety and problems. The conflicts are presented as a conflict chart
(see Fig. 4) and the risks as tables or histograms (an example is
given in table 1). Sometimes we also collect all observed conflicts on
a video tape which we send to the traffic planner with the study report.

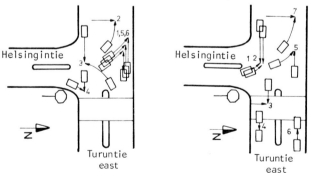

Figure 4. Conflict chart.

Table 1. Conflict and potential conflict situations and their risks
at the Helsingintie - Turuntie junction. All situations are between
two motor vehicles.

Traffic flow	Vehicles in		Flow in 8 hrs E	Risks	
	Conflict C	Pot. C P		100C/E	100P/E
Turuntie, from east					
- straight	6	26	2 500	0,24	1,04
- to left	1	5	620	0,16	0,81
Turuntie, from west					
- straight	5	37	2 864	0,18	1,29
- to right	0	4	1 044	0	0,38
Helsingintie,					
- to left	6	49	1 008	0,60	4,86
- to right	1	10	892	0,11	1,12

Training procedure

Our training program lasts for about a week. All our observers have
a driving license and most of them are students in traffic technique
so that in theory their knowledge about traffic is very good.

Day 1.: Description of the conflict technique and its uses in traffic
safety research. Presentation of the current study and its goals.
Analysis of a conflict "collection" in video and actual traffic situations in video from some previous study. The use of data collection
forms.

Day 2: Observers go through one 60 minute video tape, which is then
analysed together with the instructor. More video tapes are analysed
with the instructor.

Day 3: Observation training at a junction, first with the instructor,
then without him. Afterwards in the laboratory, the video tape of the
morning's observations is analysed.

Day 4: Observations at a junction and analysis of them in video afterwards.

Day 5: Observations at a junction and analysis of them in video afterwards.

The observers' reliability is tested during the analysis of their observations and the results of these tests are then inspected immediately with the observers.

Choice of observation periods

We nearly always make conflict observations in good weather conditions. Sometimes, when it has been of interest, we have observed conflicts in rain or on slippery road surfaes. The reason behind observing only in good conditions is that we try to eliminate the effects of different weather conditions when comparing locations of before and after situations. This also means that our observations concentrate in time periods 15.4. - 20.6. and 1.8. - 15.10.

Daily observation periods are typically 7.30 - 9, 11 - 12, 14 - 15, and 16 - 17. Sometimes we delete the 7.30 - 9 period depending on the wishes of the commissioner of the study. In total we usually make observations in one location for 6 - 12 hours. This amount has been too small for drawing significant conclusions about the locations' safety problems but we are always bound by the usually meagre recources granted for these studies. Finnish towns have very small funds for use in safety research.

Evaluation of our technique

We test the observers' reliability during the training program, as mentioned before. Some results from these tests are shown in table 2.

Table 2. The number of correctly scored conflicts (A), conflicts not observed (B), situations falsely scored as conflicts (C), and the correct score rate 100 A/(A + B + C) % for four observers during their training.

Amount of training or experience	Video tape or junction observations	A	B	C	Correct score rate (%)
1 day	video tape	16	8	18	38
4 days	junction	19	13	9	46
5 days	junction	23	4	3	77
11 days	junction	27	2	2	87

We have now had the same observers for two years, and their correct
score rate is about 80 %. The method's reliability is, however,
higher, because we use video in all studies and check all observations, both conflicts and potential conflict situations, afterwards.

We have so far had to rely on Christer Hydén's validation studies in
Lund with the same kind of technique. All our studies have been commissioned by towns or Finnish Road Administration, and we have not
been able to get funds for a sufficient validation study. After a few
years we hope to have gathered enough information from our conflict
studies to measure our method's validity in Finnish conditions.

Anticipated modifications in Malmö

The number of observers is quite normal (three) but it is highly
possible that we can't register potential conflict situations because
of the high traffic flows in the junctions. We have to concentrate on
conflicts. Otherwise we don't anticipate any major modifications.

THE TRAFFIC CONFLICT TECHNIQUE
OF THE FEDERAL REPUBLIC OF GERMANY

Heiner Erke
Abteilung für angewandte Psychologie
Technische Universität Braunschweig
D 3300 Braunschweig
Germany

Introduction

The observation of critical incidents in traffic as a measure of safety
has a proved tradition in Germany. HERWIG et al. registered regular and
irregular behavior of pedetrians and drivers at marked crosswalks in
order to discover proposals for constructional corrections. In further
investigations, drivers' and cyclers' violations of the right of way
were recorded and wrong direction-indications were analysed to deduce
rules of traffic guidance (HERWIG 1960, 1965). The observation-techniques
used in these studies do not qualify for general application. The methods
were not sufficiently standardized and the relation between observations
and accidents were not investigated thoroughly.

KLEBELSBERG & SCHIBALSKI (1976) tried to define critical features in
traffic behavior at crossings in relation to safety or accidents.

The systematic development of traffic conflicts technique (TCT) started
in 1973 in Braunschweig when a group of traffic engineers and psychologists (ERKE, SCHWERDTFEGER, ZIMOLONG) analysed operational problems at
intersections.

The further development was sponsored by the Bundesanstalt für Straßenwesen, Köln, since 1975.

The development was based on general definitions and observation procedures following PERKINS & HARRIS (1968) and SPICER (1971, 1973). The observations that there are marked differences in behaviour (speed, acceleration, change in direction) and accidents (number, distribution, severity) between approaches and intersections led to the development of
different types of observation. Later TCT was adapted to conflicts especially involving pedestrians and cyclists. During the last years TCT
was applied to the evaluation of safety measures by different research
groups.

NATO ASI Series, Vol. F5
International Calibration Study of Traffic Conflict Techniques
Edited by E. Asmussen
© Springer-Verlag Berlin Heidelberg 1984

Definitions

A traffic conflict is an observable situation in which two or more road users approach each other in time and space to such an extent that a collission is imminent if their movements remain unchanged.

The event of a traffic conflict is indicated by a critical manoeuvre of at least one of the involved road users. Critical manoeuvres are:

- braking
- accelerating
- swerving
- stopping
- running, jumping
- combinations of these manoeuvres.

Excluded are safe encounters, where time and space are sufficient to coordinate the behaviour, and all manoeuvres corresponding to a traffic control device or to the roadway geometry.

The observation area is defined according to the roadway geometry, the task of the road users observed and the capacity of the observer. A conflict is counted when the corresponding accident would be located inside the observed area, the conflict can be initiated outside the area.

The degree of severity of a conflict is determined

- by the distance between two vehicles
- by the different speeds
- by the strength of the acceleration and deceleration

The severity of a conflict is determined by the estimated time span which is available to perform a critical driving manoeuvre. The shorter the time, the more dangerous the conflict, and the higher is the conflict degree of severity.

In order to be better able to grade the available time, driver reactions which still could be carried out in the given time are specified. Important is that this time is available and not wether the driver actually carried out these reactions.

Observed reactions:

- reaction to the conflict partner
- orientation to the overall situation and reaction to the other road users
- indication of own intention

Definition of conflict degree of severity 1 to 3

1: Controlled braking and/or swerving or accelerating and/or swerving to prevent a collision. The driver has just enough time to carry out these critical driving manoeuvres in a controlled manner. Time for orientation and indication of one's intention (signaling) is available.

2: Strong braking and/or abrupt swerving or fast acceleration and/or abrupt swerving to avoid a collision. The driver does not have enough time to perform these critical driving manoeuvre in a controlled manner. He has just enough time upon choosing the respective driving manoeuvre to consider the position of the other road users, but he is not able to indicate his own intentions.

3: Emergency braking and/or swerving in the "last second" or very strong acceleration and/or swerving in the last second. Only by a very fast reaction can the driver avoid a collision - a near accident. The driver is no longer able to consider the position of other road users upon choosing his driving manoeuvre.

There is no fixed set of times measured to collision or post encroachment times. The configurations of roadway geometry, collision courses, types of road users, velocities, breaking possibilities etc. make it very difficult to define time criteria - and to make these operational to the observers.

Types of road situations observed

Signalized intersections:
The conflicts usually are recorded with different recording sheets for the directions, right, through, left. Conflicts are registered in coded form. The traffic volumes are counted correspondingly.

Non-signalized intersections:
For intersections at roads with high traffic volumes and for intersections in residential areas conflicts are recorded with different procedures on different recording sheets.

Approaches:
Conflicts are recorded in coded form on a special recording sheet for the 3 sections of the approach.

Pedestrian conflicts:
A special observation procedure was conceived for areas with large

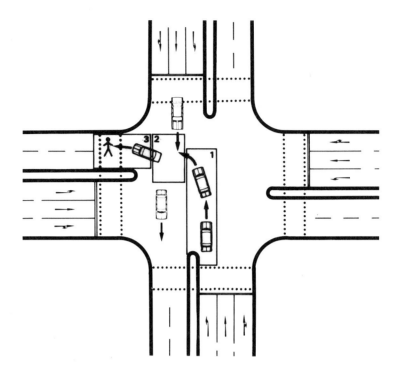

The tasks of a driver turning left

Verkehrs-Konflikt-Technik, Erhebungsbogen für den Knoteninnenbereich										P – PKW, Kombi

Ort _____ Knotenpunkt _____ Einfahrweg _____

P – PKW, Kombi
F – Fußganger
L – LKW, Transporter
M – Motorrad, Moped
R – Rad, Mofa

Datum _____ Beobachtungsrichtung: **Links**

Zeit	BEGEG	AUF	LAB	SPW	LAB-RAB	LAB-LAB	RÄUM	RÜWE	FAFÜ	FUFA

Recording sheet for the observation of vehicles turning left

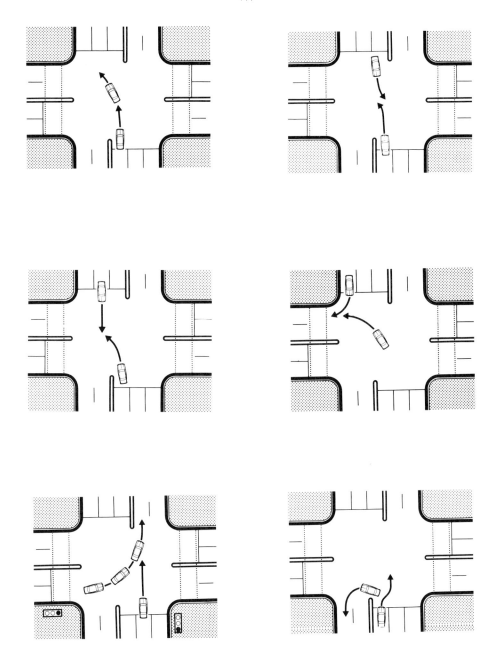

Selected conflict types for left turning vehicles

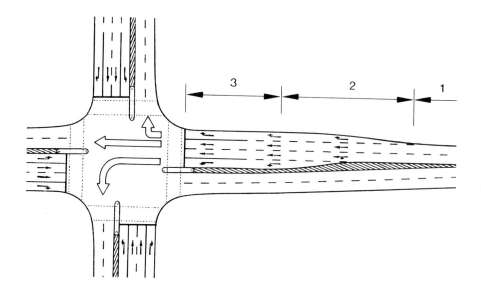

The different sections of an approach

Verkehrs-Konflikt-Technik, Erhebungsbogen für die Knotenzufahrt										

Ort _____ Knotenpunkt _____ Zufahrt _____

Datum _____ Segment _____

P – PKW, Kombi
F – Fußgänger
L – LKW, Transporter
M – Motorrad, Moped
R – Rad, Mofa

Zeit	AUF ⬆	SPW ⬆	LAB ⬆	AUS ⬆	EIN ⬆	AUS ⬆	EIN ⬆	RÜWE	FAFU	FUFA

Recording sheet for an approach

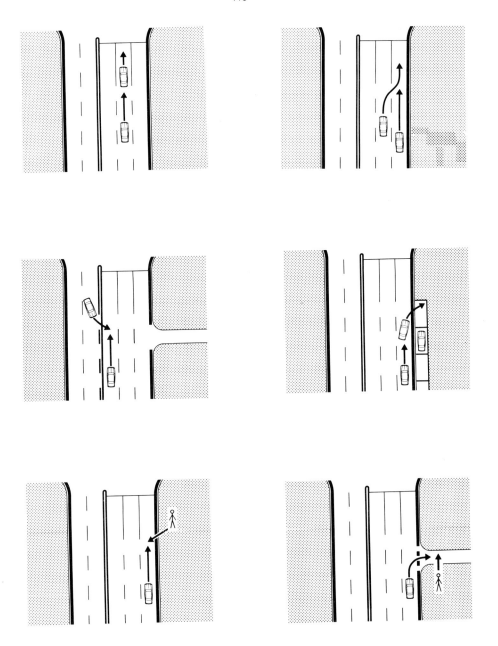

Selected conflict types for an approach

Fußgänger-Konflikt-Technik, Erhebungsbogen für LSA-geregelte Knotenpunkte

Ort _____ Knotenpunkt _____

Datum _____ Furt _____

Zeit		R	G	L	KFZ Rot	FG Rot	FG Räumen	Halten	Wenden	Innen	Außen
	K										
	E										
	A										
	G										
	K										
	E										
	A										
	G										
	K										
	E										
	A										
	G										

Recording sheet for pedestrian conflicts

Verkehrs-Konflikt-Technik, Erhebungsbogen für Konflikte, Begegnungen und Mengen

P - PKW   Z - Zweirad

Ort _____ Datum _____ Verkehrsknoten _____ L - LKW   F - Fußgänger

Zeit	Begegnungen / Konflikte					Mengen

Recording sheet for non-signalized intersections, conflicts, encounters, volumes

| VKT Fahrräder/Mofas: Konflikte, Begegnungen, Verkehrsregelübertretungen |||||

Ort _____ Datum _____ Uhrzeit von _____ bis _____

Knotenpunkt _____ Zufahrt _____

Nr.	SG	Beteiligte		Zeit

Rotlicht mißachtet   R: M:

Radw./Straße linkss.   R: M:

Radweg nicht ben.   R: M:

Gehweg befahren   R: M:

K - KFZ M - Mofa R - Fahr. F - Fußg. Konfl.:   Begeg.:

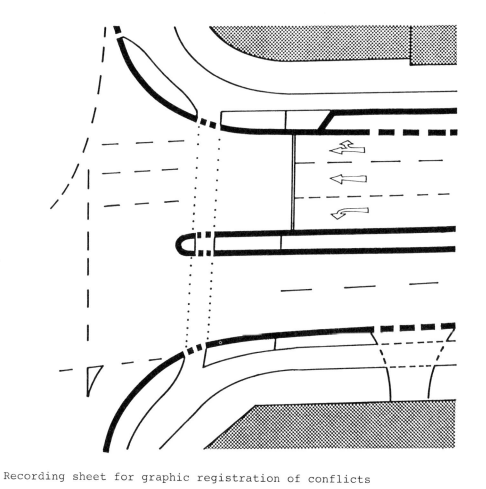

Recording sheet for graphic registration of conflicts

amounts of pedestrian traffic, intersections, zebra-crossings and road-sections outside intersections.

Cyclist conflicts:
For different locations with large amounts of cyclists a procedure to register conflicts of cyclists with vehicles and pedestrians was conceived.

Types of conflicts

For signalized intersections, approaches and pedestrian crossings different types of conflicts are defined in a verbal and in a pictorial form A simplified version of the pictorial definition is incorporated in the recording sheets together with an abbreviation of the verbal definition.

For other locations, especially for conflicts involving pedestrians and cyclists, a graphic signing procedure was conceived. The conflict is represented with arrows corresponding to the directions of the road users involved. Special types of movement are recorded by an analogous line. Braking, accelerating and evasive actions are recorded as well. The instigation and the active contribution to the compensation of the conflict situation are noted.

Data collection procedure

Trained observers, usually teams, are positioned at the location under observation to be able to detect the conflicts with a perspective similar to that of the road users involved.

For registration the observer uses the appropriate recording sheet. The traffic volumes are recorded in an analogous way.

For conflicts with left-turning vehicles involved encounters were registered in addition to the volumes.

For studies with special emphasis on pedestrian and cyclist conflicts it seems to be useful to note encounters and violations of traffic laws

At locations with low traffic volumes and few conflicts the observer ha to register traffic volumes, encounters, violations and other critical incidents in addition to the conflicts.

Data treatment procedure

At the end of every observation period data are reviewed to identify obvious errors and double recordings when two or more observers were recording. When conflicts are recorded with graphic notation they are classified corresponding to main configurations of movement and irregularities, especially for cyclists and pedestrians.

The corresponding accident data are gathered usually for the last two to four years. For validation studies usually only those accidents are selected that happened under conditions corresponding to those of the conflicts observation.

For the comparison of types of conflicts and types of accidents a classification system for accident causes was conceived. We use all types of accidents for the comparison with conflicts. The analysis of accidents at the different locations showed that the severity of accidents is not only a consequence of the features of the movement pattern common for conflicts and accidents but is also a consequence of the types and age of road users, use of seat belts etc. The relations are expressed in accident-conflict-ratios for different locations and manoeuvres. The values are computed for one year and multiplied by a factor of 10^5.

Accident conflict ratios
Signalized intersections

	Rear end	Left turn	Lane change	Vehicle pedestrian	All manoeuvre
Approach	0.41	-	0.78	-	0.49
Inner section					
Right	1.33	-	-	1.18	1.11
Through	0.72	7.75	2.81	8.32	3.04
Left	0.57	3.50	-	2.72	1.83

Nonsignalized intersections

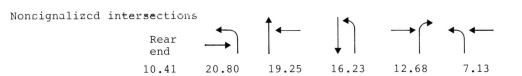

Rear end					
10.41	20.80	19.25	16.23	12.68	7.13

Training procedure

Observers are trained with a training manual and with observations at
different roadway situations. Usually the interrater reliability is
better than .70 after 2 days of training.

Choice of observation periods

For selected intersections, signalized and non-signalized, we observed
traffic conflicts during a whole week from 7 a.m. to 7 p.m. The days
from monday to friday are very similar in frequency and distribution of
conflicts, saturday and sunday are different. For special situations
conflicts were observed during the night as well - with almost no conflicts recorded.

If possible observations should cover 12 hours form 7 a.m. to 7 p.m.
When time or manpower are limited the observation should comprise at
least 6 hours, for example 7 to 9, 11 to 1, 3 to 5 or 9 to 11, 1 to 3,
5 to 7.

Usually the observation time is divided into intervals of 25 min. obser
vation and 5 min. break or 45 min. to 15 min. when observation area or
position are changed.

Evaluation

For most locations observed the reliability of the occurence of
conflicts was tested:

Situation	Hours of observation	Number of locations	Correlation
Signalized intersection	2 x 5 x 12	2	
Monday to friday Intercorrelation			.83**
Distribution During the day, split-half			(1) .77** (2) .66**
Non-signalized	1 x 5 x 12	1	
Monday to friday, split-half			.87**
Distribution During the day, split-half			.88**
Cyclists, split-half	32 x 6	32	.48**

For all situations where TCT was applied we computed the correlations
of conflicts and accidents:

Situation	Number of locations observed	Accident-conflict-correlation
Signalized intersections Approach	24	.86**
Signalized intersections Inner section (1)		
Left	24	.77**
Right	24	.29
Through	24	.23
Signalized intersections Inner section (2)	14	.78*
Signalized intersections		
Vehicle-pedestrian (1)	24	.36
Vehicle-pedestrian (2)	12	.72**
Non-signalized intersections	10	.93**
Roadsection Outside intersections Vehicle-pedestrian	6	.35
Cyclists	32	.70*

For some conditions multiple correlations show better results when
encounters and traffic volumes are considered as well.

TCT was applied to different diagnostic and evaluation tasks (ALBRECHT,
1982; HOFFMANN & ZMECK, 1982). TCT is suggested for traffic control by
the police (STASTNY, 1983).

References

Albrecht, R. Evaluation of traffic restraint measures in residential areas with respect to pedestrian safety. OECD Seminar on short-term and area-wide evaluation of safety measures. Amsterdam, SWOV 1982, 154-160.

Erke, H., Gstalter, H. & Zimolong B. Verkehrskonflikttechnik. (1) Handbuch für die Durchführung und Bewertung von Erhebungen. (2) Trainingsheft für Beobachter. Köln, Bundesanstalt für Straßenwesen, im Druck.

Erke, H. & Zimolong B. Verkehrskonflikttechnik im Innerortsbereich. Unfall- und Sicherheitsforschung im Straßenverkehr, Heft 15. Köln, Bundesanstalt für Straßenwesen 1978.

Herwig, B. Untersuchung über Fehlverhaltensweisen im öffentlichen Straßenverkehr. Bonn, Bundesminister für Verkehr 1960.

Herwig, B. Untersuchungen über das Verhalten von Kraftfahrern und Fußgängern an Zebrastreifen. "Wir und die Straße" F.A.D 25, Bonn, Bundesminister für Verkehr 1965.

Hoffmann, G. & Zmeck, D. Sicherung der Linksabbieger an Lichtsignalanlagen. Forschung Straßenbau und Straßenverkehrstechnik, Heft 358, Bonn, Bundesminister für Verkehr, 1982.

Klebelsberg, D. & Schibalski, F. Methodische Ansätze zur Erfassung des sicherheitsrelevanten Verkehrsverhaltens an Knotenpunkten. FP 7319, Köln, Bundesanstalt für Straßenwesen 1976.

Perkins, St. R. & Harris, J. J. Traffic conflict characteristics: Accident potential at intersections. Highway Research Record, 1968, 35-43.

Spicer, B. R. A pilot study of traffic conflicts at a rural dual carriage way intersection. Crowthorne, TRRL Report LR 410, 1971.

Spicer, B. R. A study of traffic conflicts at six intersections. Crowthorne, TRRL Report LR 551, 1973.

Stastny, O. Polizeiliche Verkehrsüberwachung. Stuttgart, Boorberg, 198

Zimolong, B. Verkehrskonflikttechnik - Grundlagen und Anwendungsbeispiele. Unfall- und Sicherheitsforschung Straßenverkehr, Heft 35. Köln, Bundesanstalt für Straßenwesen 1982.

Zimolong, B. Traffic conflicts: a measure of road safety. In: H.C. Foot et al. (Eds.): Road safety. Eastbourne, Praeger 1981, 35-41.

THE FRENCH CONFLICT TECHNIQUE

N. MUHLRAD
Organisme National de Sécurité Routière
BP 34 - 94114 ARCUEIL CEDEX - FRANCE

G. DUPRE
CETE de Rouen
Chemin de la Poudrière - BP 24 - 76120 GRAND QUEVILLY - FRANCE

I. Introduction

A first conflict technique was developped at ONSER from 1973 to 1977, when a training manual for observers was issued. Some reliability and validity studies were carried out in the following years. A French conflict team took part in the first international Calibration study in Rouen, 1979, the results of which were widely used in further developments of the technique.

After the Rouen experiment, it was decided to alter some parts of the conflict data collection process, in order to make the technique easier to teach, easier and less costly to use, and to ensure that results could be immediately available. Basic definitions remained unchanged. It is this "second generation" traffic conflict technique, which is described here.

II. Définitions

The basic definition of a conflict has been agreed upon by all the teams working in the field at the ICTCT workshop in Oslo, 1977 : "A conflict is an observable situation where the interaction of several road-users (or of a vehicle and the environment) would result in a collision, unless at least one of those involved takes evasive action".

Our "working definition" of a conflict remains very near this fundamental definition, and is based on the detection of the evasive action(s) : for an observer collecting data, a conflict must be recorded :
- if a perceptible evasive action is taken by at least one of the road-users on the investigated location, and if it can be assumed that there would have been a collision without it ;
- if a real collision (damage-only accident or injury-producing one) is observed on the location.

An "evasive action" is described as a discontinuity in the driving (or cycling or walking) process, that follows the occurrence of an unpredictable or surprising event. To be efficient, an evasive action must therefore be decided upon much faster

NATO ASI Series, Vol. F5
International Calibration Study of Traffic Conflict Techniques
Edited by E. Asmussen
© Springer-Verlag Berlin Heidelberg 1984

and performed more brutally than a normal driving manoeuvre or a normal movement of the road-user observed. An evasive action is, by definition, made necessary by the traffic situation at a given moment, it is not to be mixed up with some manoeuvres practised by road-users to "intimidate" others (for instance, when trying to cross a heavy traffic flow, or when trying to make a pedestrian give way) ; such behaviour may however be at the origin of a conflict, by provoking immediately afterwards an evasive action.

Conflicts are classified on a five-point "severity scale" :

1. "Light" conflict : one of the road-users involved had to face an unexpected event, but there was ample time (or distance) to avoid the collision, and the observer has been in no doubt as to the success of the action taken.

2. "Moderate" conflict : the evasive action was more urgent and, at the beginning of the situation, the observer could believe that a collision was going to occur.

3. "Serious" conflict : the evasive action was performed very brutally and only just succeeded ; until the end, the observer believed that a collision was going to take place. (Very light damage-only accidents in heavy traffic situations, such as two bumpers touching each other, are classified into this category).

4. Conflict resulting in a "light collision" : there wasn't enough time (or distance) for an evasive action, or the action wasn't sufficiently strong, and a damage-only accident couldn't be avoided.

5. Conflit resulting in a "serious collision" : the situation developped too fast, without enough time (or distance) for an evasive action to be successful (even if it was very quick and intensive), and an accident took place, with injuries or, at least, very serious material damage.

It can be seen that the degree of severity of a conflict is linked both to the swiftness and strength required for the evasive action and to the result of this action (success or collision) : abilities of the road-users to tackle a given problem are therefore accounted for, as well as characteristics of the location itself (local design and traffic).

NB. Comparisons with other techniques, performed during the Rouen experiment, have showed that all conflicts recorded according to the French working definition, including the "light" ones, could be considered in fact as serious enough to be able to relate to accidents.

Severity of conflicts (level 5 excluded) measures a proximity to a collision, but not necessarily an injury-producing one : one can see, for instance, "serious" conflicts (level 3) that would never have produced an injury, even if the evasive action had failed. In order to be able to use conflicts as a safety indicator in replacement of injury-accidents, we have therefore established a "risk-matrix", which coefficients are proportional to the probability of an injury-accident occurring in a situation similar to that of the conflict (see Appendix). Total "risk" on a location is calculated by adding individual values of risk for all conflicts observed there.

The risk matrix was calibrated on the basis of accident-statistics and conflict data. It has been made as simple as possible to ensure quick calculation and immediately available results. The main parameters in it are the type of conflict (according to trajectories and manoeuvres), the categories of road-users involved, and the type of location. Some variables, such as speed or severity of conflicts, do not explicitly appear in the risk-matrix : either because their recording can't be made accurate enough or because their discriminating power is too low, no substantial improvement could be obtained by introducing them in risk-calculation.

III. Type of locations observed

The French conflict technique was designed primarily for use in short-term evaluation of safety measures in urban areas. In theory, all urban locations can be observed with our technique. In practice, conflicts have been so far recorded only on road-junctions and mid-block pedestrian-crossings. Use of the technique has sometimes been extended to busy rural junctions.

When a junction is being investigated, all legs are observed at the same time. In special cases, for instance on very large locations or in conditions of heavy traffic with visibility problems, it may become necessary to increase the number of observers operating simultaneously.

On a given location, all the conflicts observed are recorded, whatever movements or categories of road-users they may involve.

IV. Observation means and data collection procedures

Data-collection is performed, under normal conditions, by two observers working simultaneously on a junction. Use of technical aids such as camera, video equipment, computer etc... has been abandoned for reasons of cost, limitations to the observation-field and length of the necessary data-treatment. Video-films, however, are still being used for training purposes, and future applications of the conflict technique (information, education, etc...) may make it necessary to record conflicts in a more visual way.

As it is now, the two investigators on the ground must observe, between them, the whole location and fill one data-sheet for each conflict. The quantity of data to be filled in has been made as small as possible, in order to avoid observers missing conflicts while writing down ; at the same time, the data collected should still be descriptive enough for a diagnosis of the local situation to be possible.

For each conflict, the following items are recorded on the spot :

- name of the observer.
- precise time of the conflict.
- severity level.
- road-users involved.
- type of conflict according to manoeuvres and trajectories (see Appendix).

A quick sketch of the situation is drawn, using pre-arranged signs for some frequent actions (braking, swerving, skidding, etc...). Some brief comments are added by the observer when they can contribute to explain the origin of the conflict or describe the evasive manoeuvre.

At the end of each observation period, complementary data is added by the observers on each conflict sheet :

- references of the location.
- date.
- times of beginning and end of the observation period.
- atmospheric and road-surface conditions.
- local disturbances (road-works, police-signals, etc...).
- risk-values, calculated with the risk-matrix.

For every conflict detected, a data-sheet is filed, even if its risk-value shows to be zero ; this is an important point as the technique is intended to be used as an analytic tool as well as a quantitative one, and comprehensive conflict-diagrams are necessary for safety-diagnoses.

If both observers have recorded the same conflict, only one data-sheet will be kept in the file : a comparison between the two sets of data collected must therefore be carried out by the observers after each period on the location.

Data-collection is made on pre-printed data-sheets, which are always the same, save for the detailed map of the observed area, which of course varies from location to location. Use of a new data-sheet (such as in the Malmö calibration study) requires a short period of training for the observers, to ensure that, during field-work, it will be filled in as fast and accurately as the usual data-sheet.

The normal observation period on each investigated location covers 17 hours of an average week-day (Monday to Friday, no holidays included) ; it is, in practice, spread over three days, the observers being on the ground for two to three hours in a row. Such a long period has been found necessary if conflict data is to be used in replacement of injury-accident data : shortening the observation span means decreasing the validity of the technique. For some particular applications however, the length of the data-collection period can be cut down ; it is important in this case that afternoon and early evening hours (2 p.m. to 7.30 p.m.) should be covered as conflicts appear to be particularly frequent then, as well as late evening hours, when conflicts are rarer but with high risk-values (1).

V. Data-treatment procedure

Data-treatment procedure is very simple : most of the work has been carried out during each data-recording period or immediately after by the observers, and the conflict-sheets are ready for use.

For each location investigated, one can draw a conflict-diagram, and calculate a total risk-value by summing up the individual risk-values of recorded conflicts. Further data analysis can follow the same lines as usual accident-analysis.

According to the type of application of the conflict technique, it can be advisable either to work on raw-conflict data, or to use the risk-values. For evaluation-studies for instance, a variation of risk will be measured. For other studies, such as diagnosing why a traffic facility doesn't work, basic and comprehensive conflict-data may be required.

VI. Training procedure

The whole conflict technique described here rests on the accuracy of judgment of the observers as to the existence of, a) a situation where two road-users are on a collision course, and b) an evasive action. The main difficulties are in making the difference between an emergency action and normal traffic behaviour, and in assessing the severity of the conflict ; this second point is not very important in

---

(1) This comment refers to the average situation in French cities and important hours might vary according to the location.

evaluation studies as severity isn't explicitly part of the risk-matrix.

An intensive evasive action is easily detectable by a non-trained observer. It is less easy to decide whether an observed event is a "light" conflict or a normal traffic situation where only expected manoeuvres are performed. It is necessary therefore to "calibrate" the judgment of the observers, to make sure of the consistency of the results obtained.

A normal training program for observers speads over three days, and includes theoretical and field-work. Its content can be summarized as follows :

1st day : the aims of the technique are described. Definition of a conflict, an evasive action. Introduction of the severity scale. Examples on video-films, pictures, etc... First observation-period on a site chosen for its mixed traffic-composition and high number of light-conflicts : detection of conflicts, what is a conflict, what is normal behaviour.

2nd day : definition of "road-users involved". Origin of a conflict. Description of possible conflicts and evasive manoeuvres. Training of the observers at describing conflicts (origin, type and intensity of evasive action, consequences). Examples of light, moderate and serious conflicts on video. First introduction of the data-sheet and prearranged signs for use in sketching the conflicts. Second observation-period on a site where traffic is a bit lower than on the first day : detection of conflicts, severity, sketch and relevant comments.

3rd day : training at filling in the whole data-sheet. Exercises in the lab (video-films, etc...). More examples of conflicts for detection and severity scaling purposes. Introduction of the risk-matrix. Short training at risk-calculation. Introduction of team-work : arranging conflict-sheets according to observation-times at the end of the data-collection period, eliminating data sheets in surplus when the same conflict has been recorded by two observers, etc... Choice of position of the different observers on location to ensure good coverage. Third observation-period on the sites already investigated : detection and full recording of conflicts. Team work. Risk calculation.

During this three-day period, observers are never left alone to collect data : all their questions have to be answered, their mistakes immediately corrected.

After this training period, a few days of observation under normal conditions and on different types of locations are required, in order to familiarize the team with the variety of situations that can be found, and also to test the consistency of the data-collection and the level of reliability reached by the observers. After these tests, some observers might prove inadequate for this sort of work, some others might need an additional training period.

It has also been showed that, when conflict-observers have been working in the field for a long time (several months) training periods and tests may again be needed, to ensure that detection of conflicts is consistent, and new observers' judgment calibrates with old observers' one.

VII. Test of the technique

The first conflict technique developped at ONSER was tested both for reliability and validity. Some of the results should still apply to the new technique described here, which hasn't yet been widely used.

Reliability of observers was showed to reach a satisfactory level, provided the training procedure was properly applied, and none of the three important parts in it (indoor-teaching, guided observations, and team-discussions) were omitted. However, it was suspected that observers tend to "wear out" and should not therefore be kept on the job too long, at least not without periodical training sessions.

An attempt at validating the technique was made from data collected on twenty intersections, a good part of them being accident black-spots. Use of the risk-matrix was showed to greatly improve the correspondance between conflicts and injury-accidents, although a precise relation between these two variables couldn't be established. It was assumed that such a relation must differ according to various types of road-junctions, and this point was taken into account when designing the new risk-matrix.

Informative value of conflicts for safety diagnoses was checked on six accident black-spots. It appeared that, in such a situation, it was more difficult to detect precise safety problems from conflict data than from the direct analysis of police-accident-reports (when available). Conflicts could however be used as a complement to accident data for the design of safety measures as an indicator of various traffic problems, including minor collisions.

Use of conflict data for diagnosis purposes has still to be investigated in cases where accident data is either scarce or not available. The most important potential application of the conflict technique remains, however, the short-term before-and-after evaluation of safety measures.

A last research carried out on eleven urban sites, using the "second generation" ONSER technique, showed a correspondance between the number of conflicts and the subjective feeling of unsafety, as expressed by road-users interviewed on the sites. "Light" conflicts were more noticed and appeared more important than high-risk ones or accidents. It was assumed that low-risk conflicts were possibly related to damage-only collisions, but this couldn't be verified for lack of data.

VIII. Some practical applications of the conflict technique

The French TCT has been used several times since 1979 at the CETE of Rouen. Here are some examples of applications :

1. When traffic lights are not normally working, being either switched off or turned to flashing amber, it is the priority and give-way signs, placed on the various approaches of the junction, that tell the road-users which behaviour they must adopt. An investigation, carried out in 1976, showed that, in fact, road-users did not understand the purpose of these road-signs : in particular, 25 % of drivers on the secondary road still believed that the rule of priority to the right applied. In addition, accidents were more frequent when traffic lights were off or blinking.

An information campaign was then staged on television. After that, in 1981, an assessment of the situation was carried out and the traffic conflict technique was one of the tools used. Is was showed that a good number of crossing conflicts remained and that the campaign had not been successful in solving the problem. The effects of putting traffic lights off normal use during low-traffic hours remained desastrous.

This study showed that, when traffic lights are working, crossing conflicts are normally very rare and observation periods must be long and cover night hours. To help with this practical problem, the CETE of Rouen is now experimenting a detector, to be located on two perpendicular legs of the junction and show simultaneous approaches of vehicles from the two directions.

2. A study was undertaken to acquire a better knowledge of pedestrian behaviour at light controlled crossings, in order to improve pedestrian facilities. Again, the TCT was one of the tools used. Conflicts were analysed as to when, how, where and why they occurred. Two observers collected data simultaneously on each zebra crossing.

This study provided indications about existing facilities, such as pedestrian lights, intensity of signal, phasing of traffic lights, and so on.

3. Two types of light-controlled junctions, located on rural roads, were compared in view of measuring their efficiency. The first one featured a classical adaptive traffic-light system, the second a more sophisticated one, designed to take into account the approach speed of vehicles in order to avoid emergency braking and rear-end

collisions. The TCT was used for the comparison ; however, the number of serious conflicts collected was too small to make a firm conclusion possible.

It is to be noted that, for this particular study, an attempt was made at using only the very serious conflicts (severity 3 in the French scale). This method didn't prove operational.

Future applications of the French TCT are being considered, especially in the framework of a new safety project, now being launched by the ministry of Transports, with the aim of making the public more conscious of traffic safety. In this project, every serious accident must be investigated by an inquiry board including doctors, policemen, and road engineers ; the results obtained should be used to inform the public. At the CETE of Rouen, it is planned to use traffic conflicts in complement to accidents for information purposes : it is believed that pictures of conflicts will be more easily accepted and will therefore have more impact than pictures of accidents, and will also be much easier to shoot.

VIII. The Malmö experiment

Conditions of observation during the Malmö study should be very near our usual conditions for data-recording. However, the data-sheet is different and will take some getting used to.

Observation periods on each site will cover hours between 6 a.m. and 8 p.m., which is shorter than our normal data-collection time and will limit the validity of our results with reference to injury accidents (we usually take into account all accidents, including those occurring at week-ends or during night-hours). However, this is not an important point for a calibration study.

Altogether, the Malmö experiment should require very little adaptation of our technique.

Previous publications

G. MALATERRE, N. MUHLRAD, 1977, A conflict technique ; Proceedings of the First Workshop on Traffic Conflicts, TØI, Oslo.

G. MALATERRE, N. MUHLRAD, 1978, Training manual for the ONSER traffic conflict technique, ONSER.

G. MALATERRE, N. MUHLRAD, 1979, Conflicts and accidents as tools for a safety diagnosis ; Proceedings of the second traffic conflicts technique Workshop, TRRL, Crowthorne.

G. MALATERRE, N. MUHLRAD, 1979, International comparative study on traffic conflict technique ; Proceedings of the second traffic conflicts Workshop, TRRL, Crowthorne.

N. MUHLRAD, 1982, Les conflits de trafic, un outil d'évaluation des mesures de sécurité en agglomération, ONSER.

N. MUHLRAD, 1982, The French conflict technique, a state-of-the-act report ; Proceedings of the third international workshop on traffic conflict techniques. SWOV, Leidschendam.

APPENDIX

I. RISK MATRIX FOR A NON-CONTROLLED INTERSECTION

type of conflict	road-users involved	cars only	at least one heavy-weight vehicle	at least one two-wheeler (but no pedestrian	at least one pedestrian
11	→→ →→	1	1	2	
12					
13					
14					
15	→→ ←←	3	3	6	
16		3	3	6	
21					
22					
23		2	2	4	
24		3	3	6	
31					
32					
33		2	2	4	
34		3	3	6	
35		3	3	6	
36					
37					
41					
42					
43					
44					
45					
51					1
52					
53					2
54					
61 to 69 others					

II. RISK MATRIX FOR A LIGHT-CONTROLLED INTERSECTION

type of conflict / road-users involved	cars only	at least one heavy-weight vehicle	at least one two-wheeler (but no pedestrian)	at least one pedestrian
11	5	5	10	
12				
13				
14				
15	5	5	1O	
16	5	5	10	
21				
22	1	1	2	
23	1	1	2	
24	5	5	10	
31				
32	1	1	2	
33	1	1	2	
34	5	5	10	
35	5	5	10	
36				
37				
41				
42				
43				
44				
45				
51				1
52				
53				2
54				
61 to 69 others				

THE SWEDISH TRAFFIC-CONFLICTS TECHNIQUE

C. Hydén and L. Linderholm
Department of Traffic Planning and Engineering
Lund Institute of Technology
Box 725
220 07 Lund, SWEDEN

1.   Background

Work with developing a traffic-conflicts technique started at our department in 1973 and a technique for operational use was specified in 1974. Since then the technique has been modified and is still under further development, but many of the bases are unchanged, such as the basic hypothesis which says that there is a distinct relation between conflicts with a certain degree of seriousness and accidents.

Below the different phases of the development of the technique will be described.

2.   The original technique

The following definition was used: A serious conflict occurs when two road-users are involved in a conflict-situation where a collision would have occured within 1.5 seconds if both road-users involved had continued with unchanged speeds and directions. The time is calculated from the moment one of the road-users starts braking or swerving to avoid the collision.

The recording of conflicts was and still is made by observers at the traffic site. Tests show that observers, after approximately five days of training, are able to recognise serious conflicts with a large degree of certainty.

To analyze the relations between accidents and serious conflicts, studies were made in a total of 115 intersections in three stages:

1. Malmoe 1974-75, 50 intersections
2. Malmoe 1976   , 15 intersections
3. Stockholm 1976, 50 intersections.

NATO ASI Series, Vol. F5
International Calibration Study of Traffic Conflict Techniques
Edited by E. Asmussen
© Springer-Verlag Berlin Heidelberg 1984

At each intersection, conflicts were recorded during approximately seven hours and compared with accidents of personal injury during seven to eight years.

Analysis showed that, out of many factors, two had a definite influence upon the relation between police reported accidents with injury and serious conflicts, namely the kind of road-user involved and the general speed level at the intersection. The following average connections were obtained between the number of police reported accidents with injury and the number of serious conflicts during the same period of time.

TABLE 1: ORIGINAL CONVERSION FACTORS BETWEEN SERIOUS CONFLICTS AND INJURY ACCIDENTS

| | ROAD-USERS | |
SPEED LEVEL	CAR-CAR[1]	CAR-PEDESTRIAN CAR-BICYCLE
LOW-SPEED SITUATIONS, i.e. situations with turning motorvehicles involved, and situations with straight-on driving motorvehicles in low-speed intersections (non-signalized, mean speed < 30 km/h from all accesses)	3,2 $(2,2-5,1)$[2]	14,5 $(12,2-17,4)$[2]
HIGH-SPEED SITUATIONS, i.e. situations with straight-on driving motor-vehicles in signalized- and highspeed intersections (non signalized with a mean speed $\geq$ 30 km/h from at least one access)	13,2 $(11,2-15,1)$[2]	77,2 $(64,8-91,9)$[2]

Attention! All values in the table should be multiplied by the factor 10^{-5}.

1) The concept "car" includes lorries and buses.
2) Confidence interval with 90 % degree of confidence.

3. Second generation of conversion factors

The original technique, as described above, proved to work fairly well in operation but has had some weaknesses. Therefore the technique was slightly changed in order to give better accident-risk-predictions.

The most important weaknesses of the original technique were:

1) that the method did not give satisfying results for predicting accident risks in car-car situations, when dividing these into different types of accidents

2) that the predicted risk of an accident for two identical conflicts
could be different in different types of intersections.

To solve weakness 1) analyses were made that showed that car to car
situations should be divided into at least two groups, considering the
risk of injury, namely

A) situations where the angle between the directions of the involved
cars is less than 90°. These situations are symbolized with "Car-Car
//".

B) situations where the angle is equal to or greater than 90°. These situations are symbolized with "Car-Car ⊥ ".

Conflicts of type A turned out to be approximately four times as frequent
as type B per reported accident with injury of the same type.

Weakness 2) that two identical conflicts lead to injury accidents with
various probability depending on where they occur have been solved by
developing a new model for calculation of risk. The new model is built
on the assumption that serious conflicts can lead to personal injuries
with different probability, depending on their degree of seriousness.
Because of the basic structure of the data a division of conflicts has
only been made into two classes of seriousness. A new method of calculation has also been used to determine the connection between police-reported accidents with injuries and serious conflicts.

The following average conversion factors were received between the number
of injury-accidents.

TABLE 2: SECOND GENERATION OF CONVERSION FACTORS BETWEEN SERIOUS CONF-
LICTS AND INJURY ACCIDENTS

SITUATION CONFLICT	CAR-CAR //	CAR-CAR ⊥	CAR-PEDESTRIAN CAR-BICYCLE
Class 1 Speed < 35 km/h $1{,}0 \le$ TTC $\le 1{,}5$ sec	0	2,4	9,6
Class 2 Other conflicts with TTC $\le 1{,}5$ sec	2,8	11,9	33,9

Attention! All values in the table should be multiplied by the factor 10^{-5}

4.   Present work

Our experience with the old definition of a serious conflict led us after
some years to start a new developmental phase. New definitions are introduced and new field studies are made. Right now data are analysed in order
to obtain new conversion factors between conflicts and accidents.

From a theoretical point of view we believe that the best conversion factors will be obtained through a determination in two steps:

1. to determine the probability of each conflict leading to an accident
2. to determine the probability of each accident leading to personal injury.

It should be possible to describe the probability of a conflict leading
to an accident by the degree of seriousness of the conflict, which mainly
depends on speeds, time-margines and possibilities of avoiding the accident.

The factors determining the probability of an accident leading to personal injuries are mainly type of road-users involved, speed at collision
and angle of collision.

In the on-going analyses of the data most of the factors mentioned above
will be tested.

Two new definitions of serious conflicts are introduced in the recording
procedure. Both will be tested separately in the on-going validation
studies.

First alternative definition

A number of smaller studies in rural areas at intersections with varying
speed-limits showed us that a threshold-level depending on the actual
speed should be used instead of the fixed 1.5 seconds. This led us to
the following tentative definition of a serious conflict:

A conflict is serious if the time-margin that remains when the evasive
action is started is not more than the braking-time at hard braking on
slightly wet pavement plus half a second. The half of a second can be
regarded as the remaining reaction margin.

The relationship is illustrated in the figure below.

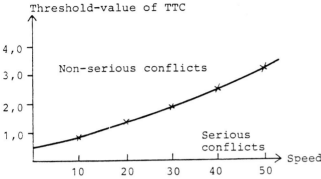

FIGURE 1: THE FIRST ALTERNATIVE DEFINITION OF A SERIOUS CONFLICT BASED
ON TIME TO COLLISION AND THE ACTUAL SPEED OF ROAD-USERS IN-
VOLVED

Second alternative definition

Earlier work has shown that there is an observable difference between
serious and non-serious conflicts. This has led us to introduce a second
alternative definition for further tests. A purely subjective severity-
scale is introduced:

Severity-rate   1 Very small risk of collision
    -"-   2
    -"-   3
    -"-   4
    -"-   5 Very high risk of collision
    -"-   6 Collision

Severity rates 1 and 2 à priori define non-serious conflicts and severity rates 3, 4 and 5 define the serious ones.

Training of observers with the new definitions of a serious conflict

The training procedure is organised in the same way as with the old definition. That means approximately five days of training mixing in-door
and out-door sessions. Video is also used as before for documentary purposes and for in-door training with edited tapes.

The first of the two alternative definitions is based on the recording of the speeds of the road-users involved and the distance to the collision-point. In both cases the recording is carried out in the moment one of the two road-users starts taking an evasive action. Training of this definition is based on estimations of speeds and distances, at first separately and then combined.

Training on the second alternative definition, the severity-scale, is based on the hypothesis that there is a distinct border between non-serious and serious conflicts (severity 1 and 2 versus 3, 4 and 5) according to the degree of suddenness and harshness introduced by the road-users involved. New observers are trained and calibrated by observers that are well-experienced in using both the old definition (with a threshold level of 1.5 seconds) and the new one.

Working procedure

The normal working procedure at the three locations chosen for the calibration study would be to use two observers located diagonally at the intersections. Parts of the centre of the intersection will be covered by both observers. After each period of observation there is a comparisc of the conflicts recorded. For those conflicts that are recorded by both observers there will be a discussion and a synthesis is made of the results by the two observers.

The normal data-sheet that is used is shown on next page. Normally only conflicts with a degree of severity of 3, 4 and 5 are recorded.

At the calibration study there will be no changes of the normal recordir procedure. The only exceptional thing will be that we will have a third observer to change with one of the two others, just for precautionary reasons.

REFERENCES

Hydén, C   A traffic-conflicts technique for determining risk.
             Department of Traffic Planning and Engineering, Lund
             Institute of Technology, Lund 1977.

Linderholm, L Vidareutveckling av konflikttekniken för riskbestämning
             i trafiken.
             (Further development of the Swedish traffic conflicts
             technique.)
             Department of Traffic Planning and Engineering, Lund
             Institute of Technology, Lund 1981.

CONFLICT RECORDING SHEET

| Observer: | Date: | Number: | Time: |

City: ....

Intersect: ....

Weather: Sunny ☐   Cloudy ☐   Rain ☐

Surface: Dry ☐   Wet ☐

Time interval: 9^{30}-10^{30} ☐ 10^{45}-11^{45} ☐ 12^{30}-13^{30} ☐ 13^{40}-14^{40} ☐ 15^{00}-16^{00} ☐ 16^{10}-17^{10} ☐   Other ....-....

north?

	Road-user I	Road-user II	Secondary involved
Private car	☐	☐	☐
Bicycle		☐	☐
Pedestrian		☐	☐
Other	....	....	....
Sex (ped)	M☐ F☐	M☐ F☐	M☐ F☐
Age (ped)	....years	....years	....years
Speed	.... km/h	.... km/h	.... km/h
Distance to coll. point	.....mtrs	.....mtrs	
Time to coll	.....sec	.....sec	

Avoiding act.	Yes	No
Braking	☐	☐
Swerving	☐	☐
Accelerating	☐	☐

Possibility to swerve	Yes ☐	Yes ☐
	No ☐	No ☐

Estimated risk of a collision:

☐ 1 very small
☐ 2
☐ 3
☐ 4
☐ 5 very high
☐ 6 collision

Description of the causes of event:

Continued on the other side: ☐

Sketch including the position of the road-users involved.
Please mark your own position with X.

☐→ Private car; Lorry; Bus

—o→ Bicycle; Motorbicycle

X→ Pedestrian

APPLICATION OF TRAFFIC-CONFLICT TECHNIQUE IN AUSTRIA

R.Risser & A.Schützenhöfer
Austrian Road Safety Board

In Austria using traffic-conflict-technique we had two aims :

A) To identify certain spots in the road-network, where the danger of accidents is high, before many accidents actually have happened.

B) To describe drivers. The question is, if the number of conflicts in which a driver is involved on a certain route can be used as an index typical for the driver.

1. Traffic conflict registration on-the-spot

Traffic conflict registration on-the-spot was mainly performed to identify spots in the road-network with a higher accident risk.

1.1. Actual traffic-conflict definition

Serious conflict : Rapid deceleration or emergency braking, rapid or violent change of direction - with a character varying from a lane change to a swerve. The time for this manoeuvre is too short to consider other vehicles or pedestrians not directly involved in the traffic-conflict.

Slight conflict : Controlled braking or change of direction is necessary to avoid a collision. There is ample time to keep control of vehicles or pedestrians not directly involved in the conflict.

Precautionary braking or change of direction to avoid a conflict is logically not defined as a conflict; which means that we would register a conflict only in the case, that an evasive action has become necessary to avoid an accident.

Following this definition we would state that situations with a very narrow escape, but without an actual necessity of an evasive action, are not to be defined as traffic-conflicts. Such situations we would rather call "near misses without evasive action" (where you never can tell if it was the skill of the driver which let him accept such a dangerous situation, or if it was nonchalance, or lack of ability and just chance that an accident did not happen).

The character of the evasive action in our definition on one hand

NATO ASI Series, Vol. F5
International Calibration Study of Traffic Conflict Techniques
Edited by E. Asmussen
© Springer-Verlag Berlin Heidelberg 1984

is a condition for the existence of a traffic-conflict, on the
other hand it is a base for judging if a conflict was a slight or
a serious one : Can the actions to avoid an accident take place
with regard to other participants in road-traffic who are not
directly involved in the conflict ? All emergency actions are
symptoms for serious traffic-conflicts.

As one can see, in our definition we did not use any objective
measure (like meters or seconds left till the imminent accident).
The reason for this is, "than man is in possess of enough ability
to recognize very complex happenings" (HÖFNER & SCHÜTZENHÖFER,
1978).

1.2. Traffic-conflict-types registered

Conflicts are described considering severity, localisation, the
involved vehicles, their position in relation to each other re.
the way they approach each other.

To make work easier observers use simplified plans of the observed
intersection. They have to record where the collision would have
taken place if no evasive action had happened (see fig. 1).

1.3. Observation means

Traffic conflicts are registered in vivo.

1.4. Data collection

As results of the traffic-conflict registrations we gain plans
containing the localisations of the traffic-conflicts and/or traff
conflict diagrams (see fig. 2).

1.5. Training of observers

Training of observers at first is done by means of video-recording
followed by supervised registration and discussion of conflicts in
vivo. Comparing local and chronological details in the registratio
of one conflict by different observers their interrater-correlatio
can be analyzed. The time it takes until a sufficient interrater
correlation is obtained differs from trainee to trainee and from
situation to situation.

1.6. Practical use of traffic-conflict registration on-the-spot

Following a pilot study in 1978 traffic-conflict technique in the
mean-time was established as a means of identification and diagnos
of black spots in the city area of Graz.

Authorities responsible for the construction of roads and bridges had a team of observers trained. Furthermore, traffic-conflict technique is part of practical courses dealing with traffic-psychology, lead by SCHÜTZENHÖFER at the University of Graz.

Graz until now is the only city in Austria whose authorities have traffic-conflicts registered as a routine job.

The traffic-conflict technique used in Graz appeared to be a valid method to identify spots with a high accident risk for bicycle riders.

One disadvantage of the registration on-the-spot lies in the fact that until now it was limited to analysis of cross-roads re. junctions. Road sections between cross-roads could not be analyzed using this technique, yet.

2. Registration of traffic-conflicts out of moving cars

In 1982 the Austrian Road Safety Board in Vienna conducted a study with the following goals :

1. Finding out typical drivers' errors resulting from unadjusted behavior and often leading to traffic-conflicts.

2. Describing and recording these conflicts and their reasons.

200 subjects were observed while driving along a standardized route in Vienna. 2 observers in the subjects' cars collected the necessary data (RISSER, TESKE, VAUGHAN & BRANDSTÄTTER, 1982).

2.1. Modified definition of traffic-conflicts

The definition used in this project shows 2 new characteristics compared with the definition given in 1.1.:

o We found out, that even acceleration can be a criterion for the existence of a traffic-conflict; and again the decision if a traffic-conflict is to be judged as a slight or a serious one depends on it the acceleration was controlled but rather powerful or if it was rather violent.

o The second difference : We tried to answer the question if the conflict was totally or partly caused by the observed traffic participant or if he had no chance to avoid the conflict at all.

2.2. Collected data and observed situations

The above given definition of traffic-conflicts does not contain

dangerous situations arising in absence of other traffic-participants. That's why in the course of the project in Vienna we
collected the special events described below in an extra category :

o Near misses in the absence of other traffic-participants, both
   when evasive action was necessary and when it was not necessary.

o Deviations from the actual road-area (e.g. driving on the banque
   driving over a traffic-island, etc.).

But events like these happened very rarely. The most important
events to describe were traffic-conflicts. Applying the traffic
conflict technique the way we did we gained possibilities to
identify and record certain interaction characteristics that
happened before the conflicts and to analyze their relations to
the actual traffic-conflicts.

Moreover, description of the traffic-conflicts did not consist in
describing the kind of approach traffic-participants took in
relation to each other but in describing the behavior of the
involved parties (e.g. "stubborn" behavior insisting on the right
of way, etc.).

2.3. Means of observation

The traffic-conflict registration in the course of the Vienna
project was done in vivo, too. The actual criteria "danger" re.
"narrowness of escape" cannot be mediated by video-recordings. On
the other hand, the video-technique is very useful as a help to
remember critical situations and to discuss them after a certain
period. Furthermore, video-recordings can help to demonstrate
observing-trainees some typical aspects of conflict registration
before one starts supervising them in vivo.
Actually, as a means of recording some typical conflicts and as
an aid in the first training stage of observing-trainees, we used
video-recordings done from a car following the subjects' vehicle.

2.4. Data collection

In the Vienna project it was impossible to design traffic-conflict
plans or diagrams. We simply counted traffic-conflicts happening
on each of the 51 sections of our test-route; besides, the observe
had to decide whether the conflict was slight or serious and wheth
it was wholly or partly caused by the subjects or if the subjects
did not have any chance to avoid the conflict.
One of the observers described the behavior of the subjects in a

standardized way along the whole route, which enabled us to analyze interaction before and during traffic-conflicts.

2.5. Data processing

In the course of our project we obtained indices for the 200 subjects observed, based on the traffic conflicts and other critical
events they were involved in driving along our test-route. Concerning the sections of the test-route we also aquired knowledge
on how many conflicts and other critical events occured on them.
These data were later on compared to the numbers of accidents which
had happened there during two different periods.

The data concerning the subjects were set in comparison to in-
formations the subjects had given us about their driving record
(e.g. accidents and fines).

2.6. Training of observers

As mentioned above the first training steps were done with help
of video-recordings. In the Vienna project trainees had to observe
15 to 20 tours on the monitor. At first critical events were
identified for them; as soon as possible they should learn to
recognize possible traffic-conflicts by themselves - although,
as already mentioned above, video-recordings of traffic-conflicts
lack important details of essential conflict criteria.

This way our trainees had to recognize and discuss between 40 to
60 conflicts on the monitor before they could try to identify
conflict situations in vivo.

We tried to obtain a high interrater-correlation when identifying
traffic-conflicts without helping the trainees by providing them
with possibilities to quantify the base of their judgement (e.g.
meters or seconds until the imminent accident).

2.7. Evaluation

To measure interrater-correlation it was not possible for 3 or
more observers to drive along the test-route in one car - this
way registrating the same conflicts - because of the observing
position affecting traffic-conflict registration. The decision,
if a situation is already to be called a conflict or is just not
a conflict yet is very much influenced by the position of an
observer. We found out, that the observer who has to register
conflicts always should sit beside the driver. The task performed

by the other observer, doing the standardized description, is not that much affected by the position.

Anyway, the difficulties arising when registrating conflicts from different positions in the car made us choose a special form of interrater-control : Supposing, that observation results should show similar distributions in different sub-samples, provided those sub-samples are large enough, we compared the results of the six observers taking part in conflict registrations in the course of our project, analyzing the 15 possible pair-comparisons between observers. CHi^2-tests of those comparisons were supposed not to show any significance. Actually, 12 of the 15 analyses proved a sufficiently high intercorrelation between observers (which amounts to 80 % of possible correspondence). Taking into consideration only conflicts wholly or partly caused by the subjects the amount of correspondence mounts up to 87 %. This means, that different ob-servers obtained similar results (which is meeting our expectations, provided that similar results are not based on different facts).

As far as validity of traffic-conflicts is concerned, we found 4 significant correlations between conflicts persons had on our test course and former accidents they were involved in : The only posi-tive correlation was found between "high-speed accidents" in ab-sence of other traffic-participants ($r = 0,155$; $t^{err} = 0,05$). Accidents with bicycle-riders and pedestrians ($r = -0,15$), lane change accidents ($r = -0,14$) and overtaking accidents ($r = -0,19$) are negatively correlated to the conflicts we registered when observing our subjects on the test-route. If, instead of the correspondence between driving-records and conflicts on the test route of different persons one analyzes a correlation between con-flicts on the 51 sections of the course and accidents registered on all those sections during 2 different periods, the results are distinctly better :

    Conflicts with severe personal injuries 1981 $r = 0,73$
    Traffic conflicts with a total number of
            accidents with personal injury 1981 $r = 0,55$
    Conflicts with accidents resulting in
            material damage only 1981 $r = 0,52$

When taking into consideration accidents from 1976 to 1980 the results are similar.

Quite obviously it is different, if you have to use conflict numbers as an index for a person or as an index for a certain section in the road-network.

Maybe there is quite a simple explanation for this discrepancy :
Most of the conflicts we registered are "slow conflicts", as we
want to call them. Those persons, who got involved into conflicts
of this type quite often reported few accidents in their driving
record. Errors which are the causes of "slow conflicts" (inadequate
precautions on cross-roads, driving off at the wrong moment, hesi-
tating re. inconsequent change of the lane re. lane-choosing for
turning off, inadequate distance to pedestrians or cyclists)
obviously result in accidents quite rarely, although, in the course
of a year, the number of accidents in the road-network is high
enough. Only, when analyzing the behavior of persons, one finds
out that other types of behavior - which do not result in conflicts
this often (at least, they don't in the eyes of the observers) -
are found in persons who according to their own statements have
been involved in traffic-accidents quite often.

In connexion with those types of behavior we distinguished the
following errors :

- Errors in connection with too high a speed
- Too small distance to the preceding
- Errors defined as violations (e.g, driving over an intersection
  when the traffic-light shows "Yellow", etc.)
- "Stubborn" behavior insisting on the right of way
- Risky overtaking.

If errors of these types result in conflicts - which does not happen
very often - we call them "fast conflicts".

Maybe the statement, that those errors do not lead to conflicts
very often, is wrong : Possibly, it is very difficult for observers
to recognize slight conflicts resulting from them, because those
slight conflicts do not disturb the impression of fluency of traffic.

When describing persons by means of the errors that lead to slow
conflicts on one hand and to fast conflicts on the other hand you
can distinguish 4 types of drivers :

Type_1 :
Drivers with many conflicts and many accidents. They often commit
errors of both types.

Type_2 :
Drivers with few conflicts and many accidents. They mainly commit
errors leading to fast conflicts.

Type_3 :
Drivers with many conflicts and few accidents. They commit errors
leading to slow conflicts.

Type_4 :
Drivers with few conflicts and few accidents. They commit few erro
of both types.

2.8. Expectations for the Malmö experiment

The fact that in our last project we registered conflicts while
driving along a standardized route with the observed subjects does
not mean that we plan to do no more on-the-spot registrations. On
the contrary : For a detailed analysis and diagnosis of sections
of the road-network one cannot renounce on-the-spot registrations.

That is, why one question seems very important to us : Is a person
who has learned to register conflicts out of a moving car able to
do an adequate job when registrating conflicts on-the-spot ? Does
he obtain the same results than other observers do when recording
conflicts together with them ? The last question is : Does a conflict registration out of a moving car lead to results which are
comparable to recordings persued by observers on site ?

We will have to find out how far the position of the observers
does affect the gained results. One has to expect such positional
influences, because 2 or more observers recording conflicts posted
on different spots of one cross-road will also come to different
results. On the other hand one has to consider, how far kinetical
factors are to be taken into consideration : When an evasive
action takes place the observer can make use of additional stimuli
(beside the optical and acoustical ones) to decide, if a conflict
has taken place, namely the kinetical stimuli, which in our opinio
are distinctly of great help when registrating decelerations,
swerves or accelerations.

One observer from Austria will take part in the Malmö experiment,
the results of his traffic-conflict registrations should be compared to results of observers using a definition similar to the
Austrian one to identify conflicts.

Literature :

HÖFNER, K.J. & SCHÜTZENHÖFER,A.: Konfliktforschung im Straßenverkehr. Verkehrsjurist d. ARBÖ, Nr. 39/40, Wien 1978.

RISSER,R., TESKE,W., VAUGHAN,Ch., BRANDSTÄTTER,Ch.: Verkehrsverhalten in Konfliktsituationen. Kuratorium f. Verkehrssicherheit im Auftrag des Jubiläumsfonds der österreichischen Nationalbank, Wien, 1982.

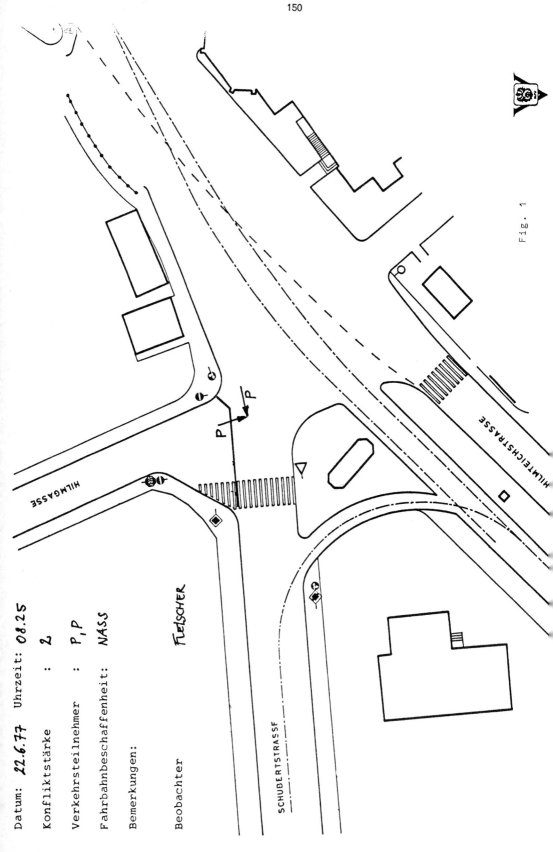

Datum: **22.6.77** Uhrzeit: **08.25**

Konfliktstärke : **2**

Verkehrsteilnehmer : *P , P*

Fahrbahnbeschaffenheit: *NASS*

Bemerkungen:

Beobachter *FLEISCHER*

SCHUBERTSTRASSE

HILMGASSE

HILMTEICHSTRASSE

Fig. 1

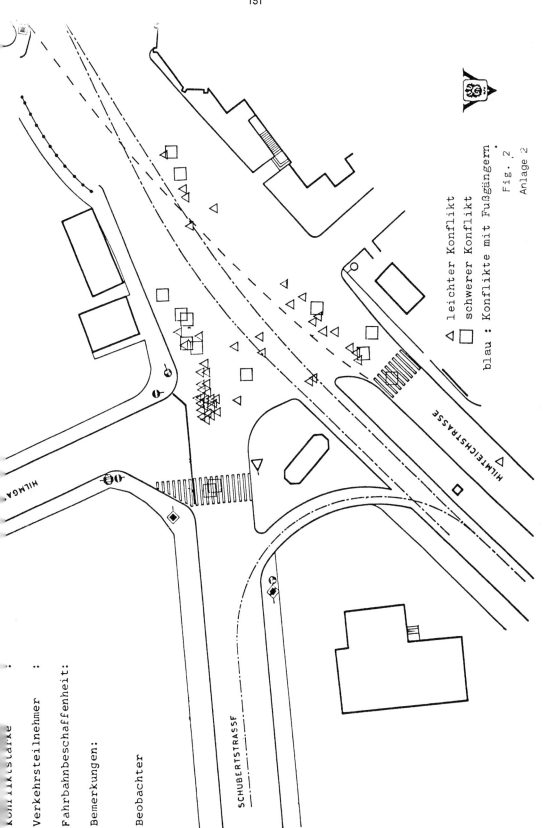

△ leichter Konflikt
□ schwerer Konflikt
blau : Konflikte mit Fußgängern

Fig. 2

Anlage 2

THE USE OF TRAFFIC-BEHAVIOUR-STUDIES IN DENMARK

Ulla Engel & Lars Thomsen
Danish Council of Road Safety Research, the Secretariat
Ermelundsvej 101, DK-2820 Gentofte, Denmark

This paper does not deal with traffic conflict techniques, but rather studies of traffic-behaviour. In the following, three types of studies are revealed and compared with the corresponding statistical accident analysis.

It should be born in mind that the studies concerned all are included in the traffic restraint programme implemented in Copenhagen, Østerbro. Some material covering this 11 year-project can be found in the papers Engel (1982), Thomsen (1982) and Engel & Thomsen (1983a, b and c).

1   Some necessary background-remarks

The Danish Council of Road Safety Research proposed in 1971 to the road authorities of the City of Copenhagen to carry out a joint research project. The aim of the project was an area-wide traffic replanning of a part of Østerbro, a residential area with 17,000 inhabitants. The aim of the scheme was to reduce the number of accidents by simple physical countermeasures.

The project consisted of three stages:

1. Collecting data and proposing a scheme
2. Implementing the scheme
3. Evaluating the traffic safety effect of the scheme

The evaluation of safety measures is primarily based on an analysis of the traffic accidents in the area, which have taken place before and after the implementation of the scheme. But also studies of the behaviour of the road users have been carried out in order to registrate whether or not the intentions of the countermeasures were obtained.

In the before-period (1969-1971) 475 police reported accidents took place, and in the after-period (1977-1980) 370 accidents took place.

NATO ASI Series, Vol. F5
International Calibration Study of Traffic Conflict Techniques
Edited by E. Asmussen
© Springer-Verlag Berlin Heidelberg 1984

However, we are still dealing with small numbers of accidents, since the traffic scheme consists of 25 different countermeasures and each of them is directed towards specific (and more or less different) accident types.

The study has tried to provide answers to the following questions:

1. Which conclusions can be drawn from the results concerning the reduced number of accidents and number of persons injured, when defining "short term" as a period of 7-10 years and "area-wide" as a sum of 25 physical countermeasures implemented in a residential area of about half a square kilometre?
2. How do we distinguish between the accident reducing effect of the implemented safety measures, the reduced number of road users in the area and general safety measures implemented in the whole country in the same period?
3. To what extent have the behavioural studies supported or invalidated the results of the accident analysis?

The proposal of the scheme was based on a very thorough and detailed accident analysis. This analysis was based on the accident reports made by the police, these being the best material for detection of the possible *causes* of the accidents. The accident analysis thus provided hypothesis concerning accident-prone behaviour of the street-users at given locations.

As might be imagined the number of accidents in the project area is relatively small when operating at a very detailed level. Furthermore there is quite a long distance of time between the before and after period. These circumstances have lead to the implementation of different studies of the road users behaviour before and after the implementation of certain countermeasures. Emphasis has been given to studies concerning the speed of motorvehicles, the lateral position of different road users in the carriageway and pedestrians use of different crossing facilities.

We can tell from these studies whether the road user changes his behaviour according to the intention laid down in the countermeasure. So the purpose of the studies was to enlarge the background for expectations regarding the effect of each countermeasure. We do not know, however, whether the behaviour of the observed normal road users is similar to the behaviour of road users involved in accidents, but further research in this type of before/after-studies might clarify these possible relations.

2   Three groups of behavioural studies

After these introductory remarks we will proceed giving a more detailed
description of three groups of traffic-behaviour-studies and relate
these results to the before/after-trends in the accident-figures.

2.1   The behaviour of car drivers and bicycle-riders related to
different street markings. Crossing-behaviour of pedestrians
in Nordre Frihavnsgade

In the streets Nordre Frihavnsgade, Randersgade and Århusgade it has
been tried to alter the behaviour of the street-users by means of different types of stripes on the road. It has been the purpose to change
speed and spacing for the drivers, while it has been the idea to provide the pedestrians with well situated locations for the crossing
of Nordre Frihavnsgade. This study is described in two working papers
(in Danish) by Engel & Thomsen (1978) and Thomsen (1982) and a thesis
by Ingason (1981). The related accident analysis will be found in
Notat 1/1983. Trafiksanering på Østerbro. Del 1 _ Ulykkesanalyse, part
4.3.22 and part 4.3.24. This publication is in Danish with an English
summary.

2.1.1   Description of the accident-problem and the countermeasures
implemented as prevention.

In this section the three streets are treated in the order: Nordre
Frihavnsgade, Århusgade and Randersgade.

In the before-period in Nordre Frihavnsgade accidents of the types
211, 240 and 260 occurred, cfr. the figure at the end of this paper
showing accident-types used by the Danish police and the national
bureau of statistics. The accidents are front to front collisions and
accidents at intersections. Based on this knowledge the proposed countermeasures were centre-street-marking stripe to obtain a strict definition of the two street-lanes.

Also occurring in the before-period were accidents of the types 741,
811, 832 and 871. These were accidents with open car-doors and street-
crossing pedestrians. In order to prevent these types of accidents,
striped, double parking-places were implemented with space between

each double place. There was also implemented refuges for the pedestrians. These countermeasures were intended to provide pedestrians with
well-suited places for the crossing of Nordre Frihavnsgade. The implementation of the striped parking-places was together with the centre-markin
intended as a general "tidying up" the view of the street-users. The dri
vers were supposed to choose a more purposeful lateral placing in the
street, i.e. car drivers were intended to move towards the centre-markin
of the street without going across the centre, while parked cars were in
tended to move closer to the pavement. These changes should leave more
space for bicycle- and moped-riders.

The implemented parking-places have the measures of 5.0 m x 2.0 metres.
It was from other view-points not desirable to make them narrower than
this.

The accidents in Århusgade were in the before-period of the types 160,
240, 322, 710, 741 and 832, cfr. figure 8. These are basically front
to front collisions, accidents at intersections and accidents with
pedestrians. This street was considered as having identical accident-
problems with Nordre Frihavnsgade thus the same countermeasures were
implemented in Århusgade.

The accident-problems in Randersgade were of a somewhat other structure, especially because only one accident of the 322-type had occurred.
The general countermeasure was for this reason intended to be of speed-
reducing nature. In the west-side of the street there was introduced
parking places of the same type as in Nordre Frihavnsgade and Århusgade. In the east-side of the street parking was prohibited. In general the countermeasure narrowed the street and should in this way
slow the speed of the cars.

2.1.2   Description of the collection of behavioural data

Three aspects concerning the behaviour of the street-users have been
collected in terms of evaluation of the countermeasures. The key-words
are the speed of the cars, the lateral placing of the vehicles in the
street and the street-crossing of the pedestrians.

The speeds have been measured by means of the Doppler-effect-radar
(Engel & Thomsen, 1979) from a car parked as the "average-car", cfr.

the following section. On the spot where the vehicles were actually speed-measured another "average-distance-car" was placed in order to have comparable situations.

These measurements were carried out at four locations: "Nordre Frihavnsgade, Århusgade, Randersgade north of Århusgade and Randersgade south of Århusgade.

The lateral placing of the vehicles were recorded in Nordre Frihavnsgade and Århusgade in a very simple manner. By making faint chalkstripes on the street it was by simple observation possible to classify the vehicles according their lateral placing. At the same time the speed of the cars was measured.

In Nordre Frihavnsgade observations of the crossing-behaviour of the pedestrians were made by simply drawing on paper the routes of the pedestrians.

A view of the experimental design according to the speed measurements are given in table 1.

Table 1. Time and location of speed-measurements carried out in Øster-
         bro.

street	date	time	number of speed-measured vehicles
Nordre Frihavnsgade	thu. 10. Jun. 1976	7.30-21.15	506
	thu. 7. Jul. 1977	7.30-21.45	658
	thu. 31. Jul. 1980	10.00-19.00	299*
Århusgade	tue. 9. Jun. 1976	7.30-21.15	265
	tue. 5. Jul. 1977	7.30-21.45	378
	tue. 19. Jul. 1980	10.00-19.15	164
Randersgade 1+2 north	tue. 8. Jun. 1976	7.30-21.15	252
	tue. 12. Jul. 1977	10.00-19.15	264
	tue. 5. Aug. 1980	10.00-19.15	167
Randersgade 3+4 south	thu. 24. Jun. 1976	7.30-21.15	240
	thu. 14. Jul. 1977	7.30-21.15	475
	thu. 7. aug. 1980	10.00-19.15	159

* It is important ot note that the 1980-recordings only had a duration of two thirds of the recordings in 1976 and 1977. Important to note is also that some measurements were made in the summer-holiday while others were not in this period.

The 12 sums of <u>table 1</u> refer to the total numbers of speed-measured cars disregarding their direction, vehicle-type and traffic-situation. The total number of these vehicles is 3827.

The location of the speed measurements were chosen in such a way that the speeds measured in the two directions were supposed to be identical. The vehicle-types consist of two categories: Passenger-cars and "other" while the traffic-situation is made of the categories: Driving free, in queue and changing speed (in queue as well as not in queue).

In Nordre Frihavnsgade and Århusgade is in the years 1976 and 1977 simultaneously with the speed-measurements made a recording of the lateral position of the cars, mopeds and bicycles. For these purposes two persons were necessary.

The study of the crossing behaviour of pedestrians in Nordre Frihavnsgade was carried out in the piece of street from Grenågade to Randersgade. To make this stretch reasonable to cope with for one observer it was divided into an eastern and a western part, cfr. <u>table 2</u>.

<u>Table 2</u>. Design for studies of crossing-behaviour of pedestrians in Nordre Frihavnsgade between Grenårgade and Randersgade including the intersection at Randersgade.

location	date	time	number of observations
easter part	9. Apr. 1981	12.00-12.26	
		14.19-15.12	401
western part	21. Apr. 1981	16.39-17.33	

The pedestrians were divided into groups according to age (three groups) and the use of the corner-refuges (yes/no).

2.1.3   Results with relation to road-markings and corner-refuges

The results are treated in four groups: The speed of the cars, the lateral placing of the vehicles in the street, the connection between speed and lateral position and finally the crossing behaviour of the pedestrians.

It should be born in mind that the figures given in table 3 are for the
total number of speed-measured vehicles. The mean-speeds are seen to
be between 42 and 47 km/h while the standard deviation is between 7
and 9 km/h.

Table 3. Results from the speed-measurements made at the four locations.
For each combination of year and street mean speed, standard
deviation and numbers of observations are given. Speed-figures
are given in K.p.h.

year/ location	1976 before	1977 after	1980 after
Randersgade, 1+2, north	41.9	44.5	42.6
	7.4	8.7	8.4
	252	264	167
Randersgade, 3+4, south	45.3	46.6	45.8
	8.5	9.3	9.2
	240	475	159
Århusgade	44.4	43.2	41.9
	7.9	9.1	9.0
	265	378	164
Nordre Frihavnsgade	46.7	46.2	42.8
	7.6	8.9	9.3
	506	658	299

The first comparison concerns the before/after-situation, i.e. 1976
compared with 1977.

It can be found that there is an increase in the standard-deviation
from 1976 to 1977. The general situation is that there is no change
in mean-speeds from 1976 to 1977. Analysing the streets separately it
is seen that the northern part of Randersgade has an increeasing mean-
speed while Århusgade shows a tendency towards a decrease in mean-speed
from before to after the implementation of the changes.

These results are found for all groups whatever vehicle-type or traffic-situation.

As a kind of control speed-measurements were carried out in 1980 as
well. In terms of countermeasures the situations of 1977 and 1980 were
identical. Comparing the three years one finds a decreasing mean-speed
from 1977 to 1980, while the standard deviations of 1977 and 1980 are
above those of 1976.

The parking places have not had the intended effect as the parked cars
have moved in average 7 cm closer to the centre of the street, cfr.
table 4. The change may have occurred because drivers now aim at the
left side of the parking-place.

Table 4. Kerb-distance characteristics of parked cars. All distances
        are in metres.

time/	1976 = before			1977 = after		
location	member of observations	mean distance	standard deviation	member of observations	mean distance	standard deviation
Ndr. Frihavnsgade	401	0.18	0,12	332	0.26	0.13
Århusgade	569	0.20	0.14	395	0.26	0.14

The distance from the driving vehicles to the pavement is unchanged in
Nordre Frihavnsgade while it has decreased in Århusgade. It should
be noted that the centre marking in Nordre Frihavnsgade was introduced
before 1976, which means that the 1976-situation in Nordre Frihavnsgade
is not the genuine before-situation.

In both streets the standard deviation of the distance between the
driving cars and the pavement has decreased in both streets. This seems
reasonable as the parked cars have narrowed the street. The distances
for the driving vehicles are given in table 5.

The same kind of figures for bicycles and mopeds are given in table 6.

Table 5. Kerb-distance of driving cars. All distances are in metres.

time/	1976 = before			1977 = after		
location	member of observations	mean distance	standard deviation	member of observations	mean distance	standard deviation
Ndr. Frihavnsgade	521	3.56	0.35	507	3.54	0.31
Århusgade	280	4.01	0.42	308	3.88	0.35

Table 6. Light two-wheelers and their distance from the right-side kerb of the street. All distances are in metres.

time/	1976 = before			1977 = after		
location	member of observations	mean distance	standard deviation	member of observations	mean distance	standard deviation
Ndr. Frihavnsgade	522	2.78	0.39	569	2.87	0.36
Århusgade	334	3.13	0.47	264	3.06	0.38

The distance from bicycle-riders and moped-riders to the pavement is seen to be in the interval of 2.7 and 3.2 metres. Generally speaking there are decreasing standard deviations from before to after, while the mean-distance has increased in Nordre Frihavnsgade and decreased in Århusgade, cfr. table 6.

Because of measuring techniques the operator of the radar was instructed not to do recording while more than one vehicle was in the field of measurement. For this reason there has not been made corresponding recordings of the placing of cars and two-wheelers.

It is *probable* that the car-drivers tend to keep a constant distance from the pavement to the right and thus the mean distance between the light two-wheelers and the cars become important although the recordings are not done simultaneously.

Tabel 7 is constructed based on table 4, 5 and 6. The table shows for both years three distances: Between parked cars and driving two-wheelers between driving two-wheelers and driving cars and finally between parked and driving cars.

Table 7. Distances between parked cars, driving cars and driving, light two-wheelers. Mean-distance and standard deviation is given in metres.

period/ location	distance between parked car and driving, light two-wheeler				distance between driving two-wheeler and driving car				distance between ked car and driv. car		
	mean distance	standard deviation			mean distance	standard deviation			mean distance	standa: deviat	
	1976-1977	1976-1977			1976-1977	1976-1977			1976-1977	1976-1	
Ndr. Frihavnsgade	1.00 1.01	0.41 0.39			9.78 0.67	0.52 0.48			1.78 1.68	0.37 0	
Århusgade	1.33 1.20	0.47 0.40			0.88 0.82	0.63 0.51			2.21 2.01	0.35 0	

The mean-values seems either unchanged or decreasing. The maximum decrease is found in Århusgade where the distance between parked and driving cars has dropped 19 cm from 2.21 metres to 2.02 metres. Table 8 gives the results of the statistical hypothesis-testing.

Table 8. Results of statistical tests performed on distance-means and standard deviations comparing figures of before and after. The figures used are found in table 7.

location	distance between parked car and driving, light two-wheeler		distance between driving two-wheeler and driving car		distance betweer ked car and driv car	
	mean distance	standard deviation	mean distance	standard deviation	mean distance	standa devia⁻
Ndr. Frihavnsgade	unchanged	1976>1977	1976>1977	1976>1977	1976>1977	uncha
Århusgade	1976>1977	1976>1977	unchanged	1976>1977	1976>1977	1976>

Four out of the six mean-values show a decreased value from before to after. For the two-wheelers it is seen that in all four cases the standard deviation has decreased. The standard deviation of the distance between driving and parked cars does not show systematic changes from before to after, cfr. table 8.

For the free-driving passenger-cars there has been made a regression-analysis of lateral position and the speed. Table 9 shows sample sizes and the estimates based on the regression analyses.

Table 9. Number of observations, means, regression-coefficients and
         variance related to the four regression-analyses. Dependent
         variable is car-speed, while independent is the lateral kerb-
         distances of the car.

period/ locations		1976 = before	1977 = after
Ndr. Frihavnsgade	number of obs.	381	502
	mean-speed	46.2 k.p.h.	47.0 k.p.h
	slope	3.6 k.p.h/m	3.1 k.p.h/m
	variance	52.5 $(k.p.h.)^2$	67.1 $(k.p.h.)^2$
Århusgade	number of obs.	210	293
	mean-speed	44.6 k.p.h.	45.1 k.p.h.
	slope	2.2 k.p.h./m	4.0 k.p.h./m
	varians	51.7 $(k.p.h.)^2$	67.2 $(k.p.h.)^2$

In all but the case of Århusgade, 1976, the hypothesis of linearity
between distance and speed can be accepted. Table 10 summarises the
results.

Table 10. Results of the analysis of the relationship between lateral
          distance and speed of the cars. The table compares the before-
          and the after-situation.

location	mean speed	standard deviation	parallelism
Ndr. Frihavnsgade	unchanged	1976 < 1977	yes
Århusgade	unchanged	unchanged	yes

Except for the increase of the standard deviation of Nordre Frihavnsgade the situation is rather stable.

Ingason (1981) has analysed pedestrians and their street-crossing behaviour in Nordre Frihavnsgade. He concludes, page 46, that the refuges
and gaps between the parking places are not used more than any other
location usable for crossing. The countermeasures have thus not
changed the situation.

2.1.4   Comparing the behavioural and the accident analysis results

One of the results of the accident-analysis is given in table 11 below.

Table 11. Accidents in the streets of Nordre Frihavnsgade and Århusgade
grouped according to accident-type and period. The countermeasure evaluated is the centre road-marking found in the after-period.

accident-type period	140*	160	211	240	322	620	total
before	0	0	2	2	0	1	5
after	1	3	0	0	1	1	6

*At the end of this paper is a figure showing the full system of accident-
types.

At the top are given the accident types, cfr. figure 8 at the end of
this paper. The total number of accidents is seen to be unchanged, but
a *tendency* towards different trends from before to after for the different accident-groups is discovered. Largest change is seen in group
160 from nil to three accidents. This group involves bicycle-riders
being squeezed by cars. It is seen that this result is in good concordance with the results envisaged in table 7 and 8; the result being
that less space has been left for the two-wheeler-riders. Generally
speaking no changes can be seen neither in behaviour nor in accident-
figures.

The results of the accident-analyses can be found in sections 4.3.22
and 4.3.24 of Notat 1/1983, Trafiksanering på Østerbro, part 1, Ulykkesanalyse.

2.2   The speed of cars approaching intersections with humps

The second group of behavioural studies treats the speeding behaviour
of car-drivers approaching intersections where the car-drivers are ob-
liged to give way to the street-users on the other street. At a total

Table 12. List of the 38 intersections implemented in the after-period with hump and corner-refuges. * designates that at this location speed-profile-measurements have been carried out.

primary street	secondary street
Østerbrogade	I.E. Ohlsens Gade Carl Johans Gade Gustav Adolfs Gade Kt. Jakobs Gade Viborggade Ålborggade Århusgade Koldinggade Urbansgade Jacob Erlandsens Gade Marstransgade
Strandboulevarden	Kertemindegade Bogensegade Assensgade Fåborggade Svendborggade, north Svendborggade, south
Nordre Frihavnsgade	*Faksegade I.E. Ohlsens Gade Petersborgvej A.L. Drewsens Gade Grenågade *Hobrogade A.F. Kriegersvej
Randersgade	Bogensegade *Vordingborggade, east Vordingborggade, west Korsørgade *Koldinggade, east Ålborggade Viborggade Nøjsomhedsvej Krausesvej *Gammel Kalkbrænderivej *Rothesgade Skt. Jakobsgade
Århusgade	*Silkeborggade, north *Silkeborggade, south

of 38 junctions at Østerbrogade, Strandboulevarden, Nordre Frihavnsgade, Århusgade and Randersgade there has been implemented a narrowing of the adjoining streets together with pavements being a bit
elevated and thus being a kind of "speed-humps", cfr. figure 2 .
Table 11 is a list of junctions having these countermeasures implemented in the after-period.

Stars in table 11 indicate at which junctions the continous speed-
measurements have been carried out. The purpose of the countermeasures
has been to slow down cars while approaching the intersections. The
study is analyzed in depth in the working papers of Engel & THomsen
(1982).

The corresponding accident-analysis is found in sections 4.3.16 in
notat 1/1983, Trafiksanering på Østerbro, part 1 - Ulykkesanalyse.

2.2.1   Description of the accident-problem and the countermeasures
         implemented as prevention.

The most common accident-type in the before-period was of the 510-
category (accident between vehicles crossing at a right angle, cfr.
figure 8 at the back). Less common were the types 160, 660 and 872
concerning "squeezing" accidents at intersections and accidents with
pedestrians.

The purpose of the implemented countermeasures has been to make the
intersections easier to discover and to decrease the speed of the
cars coming from the secondary streets. If possible it was intended
to reduce the of the cars a distance of 30 metres (apprx.) away
from the intersection. The hypothesis was that a number of type 510-
accidents occurred because drivers did not realise that they were
supposed to give way at the first-coming intersection.

The countermeasures have been described in depth in the above mentioned
working papers. This will be a brief introduction. The countermeasure
consists of two components. One of them is the narrowing of the secondary street at the intersection just before the crossing pavement,
cfr. figure 2 . The narrowing has been obtained by the use of refuges
at the pavement corners. The purpose of this component is twofold.
The former is to slow down the speed of the cars while the latter is

to avoid parked cars around the corners. The other component is that the secondary street is led across the elevated pavement of the primary street, cfr. figure 2.

It has been an important aspect in this study to investigate the influence of the slope and height of the humps at the pavements. The study of this influence is restricted to car-speeds and not to accidents. Figure 3 shows the analysed heights and slopes of the humps related to time and location.

The before/after-study concerns the comparisons between 1976 and 1977. The measurements in 1979 and 1980 were carried out to get a better study of the relationship between car-speeds and hump-dimensions.

2.2.2   Description of the collection of behavioural data

The speed-measurements have been carried out for all cars that have approached the intersection from the secondary street. Figure 2 show the actual data-recording-situation. The approaching of the cars has been filmed by means of video-equipment and supplied with verbal descriptions of the traffic situation and the direction in which the cars drove at the intersection.

The traffic-situation is a grouping describing the type and amount of traffic at the intersection. The three groups are:

Intersection is empty, intersection is crossed by a vehicle and intersection is crossed by pedestrians and possibly by a vehicle. The traffic-situations were supposed to influence the speed of the approaching cars.

1,446 speed profiles were obtained by the continuous speed-recording done by the radar equipment. Speed profiles are shown in figure 4. Some profiles were left out as they were too short or the signals were too fluctuating. These disturbances may be due to more than one vehicle in the measuring-area.

Table 13 shows time, location and the number of observations for each measuring-session.

Table 13. Listing of location, date, time, duration of measurement and number of observations for speed-profiles.

location	date	time	duration	number of ob
Gammel Kalkbrænderivej	wed. 14. jul. 1976,	8.30 - 17.00	8 1/2	41
Faksegade	thu. 15. jul. 1976,	8.30 - 17.00	8 1/2 }	104
	mon. 2. aug. 1976,	8.15 - 13.00	4 3/4	
Koldinggade	tue. 6. jul. 1976,	8.00 - 18.00	10	7
Hobrogade	fri. 23. jul. 1976,	7.30 - 17.30	10	25
Rothesgade	wed. 7. jul. 1976,	8.15 - 17.30	9 1/4 }	17
	mon. 2. aug. 1976,	14.00 - 17.30	3 1/2	
Silkeborggade, south	mon. 26. jul. 1976,	8.30 - 17.00	8 1/2	28
Silkeborggade, north	thu. 29. jul. 1976,	8.15 - 16.15	8	39
Gammel Kalkbrænderivej	mon. 11. jul. 1977,	8.15 - 17.15	9 }	71
	thu. 28. jul. 1977	14.00 - 17.00	3	
Faksegade	tue. 19. jul. 1977,	8.30 - 17.15	8 3/4 }	145
	fri. 19. jul. 1977,	8.15 - 12.00	3 3/4	
Koldinggade	mon. 4. jul. 1977,	8.15 - 17.30	9 1/4 }	20
	mon. 18. jul. 1977,	15.30 - 17.30	2	
Hobrogade	fri. 22. jul. 1977,	9.00 - 17.00	8	43
Rothesgade	tue. 5. jul. 1977,	8.15 - 17.15	9 }	
	mon. 25. jul. 1977,	13.00 - 17.30	4 1/2 }	42
	fri. 29. jul. 1977,	13.00 - 16.00	3	
Silkeborggade, south	thu. 21. jul. 1977,	14.45 - 17.30	2 3/4 }	65
	mon 25. jul. 1977	8.30 - 12.30	4	
Silkeborggade, north	wed. 20. jul. 1977,	9.00 - 17.15	8 1/4 }	64
	wed. 27. jul. 1977,	13.15 - 17.15	4	
Gammel Kalkbrænderivej	mon. 20. aug. 1979, *		7 }	102
	tue. 21. aug. 1979, *		7	
Silkeborggade	tue. 24. jul. 1979, *		7 }	
	wed. 25. jul. 1979, *		7 }	143
	mon. 30. jul. 1979, *		7	
Vordingborggade	thu. 26. jul. 1979, *		7 }	141
	fri. 27. jul. 1979, *		7	
Gammel Kalkbrænderivej	mon. 18. aug. 1980, *		7 }	53
	tue. 19. aug. 1980, *		7	
Silkeborggade	tue. 22. jul. 1980, *		7 }	
	wed. 23. jul. 1980, *		7 }	145
	mon. 28. jul. 1980, *		7	
Vordingborggade	thu. 24. jul. 1980, *		7 }	151
	fri. 25. jul. 1980, *		7	

* 8.30 - 12.00 og 13.00 - 16.30

As can be seen from table 13 the data-collection has been quite time-consuming and a relatively modest number of vehicles has been recorded.

The measurings done in 1979 and 1980 has been placed on exactly the same week-days thus eliminating variation from weeks and week-days. This was not possible for the 1976 and 1977-sessions.

The implementation of humps and refuges from 1976 to 1977 did not change the speed of the cars. For this reason it was decided to exceed a height of 10 cm and a slope of 0.2 in 1979, cfr. figure 3 .

2.3.3   Results with relation to humps and corner refuges

The geometry of the eight analysed streets is rather different and only speeds nearer than 20 metres from the intersections have been compared.

20 metres from the potential point of conflict the mean-speeds are in the interval of 24 to 33 km/h. The standard deviation is approximately 5 km/h. At the pavement, i.e. the hump, the mean-speeds are in the interval from 9 to 14 km/h, while the standard-deviation is approximately 3.5 km/h.

The maximum speed measured is found 20 metres from the intersection and reaches a value of 44.8 km/h.

Figures 4 and 5 show the mean-speed-profiles for Gl. Kalkbrænderi-vej and Silkeborggade, north. As mentioned above no significant change in mean-speds were found from 1976 to 1977.

In 1979 and 1980 higher and steeper humps were implemented. The Gl. Kalkbrænderivej-intersection had in 1979 a hump with a height of 14 cm and a slope of 0.5. The decrease in speed with this configu-ration was app. 5 km/h. The before-speed was app. 15 km/h.

The intersection at Silkeborggade was changed from 1976 to 1977, cfr. figure 3 . Nothing was changed in 1979 and 1980 and the Silkeborggade-intersection was in this manner used as a control-group.

Making assumptions regarding reaction-times of drivers and the skid-resistance-coefficient between street and tyres it has been possible to group the speeds according to their degree of careful driving. The actual limits are found in figures 4 and 5 . Due to this criteria the decreasing mean-speed is accompanied by a trend towards more careful driving. This change is only obtained when the dimensions of the humps exceed a height of 8 cm and a slope of 0.3.

2.2.4   Comparing the behavioral and the accident analysis results

One of the results of the accident analysis is given in table 14.

Table 14. Accidents at the 38 intersections listed in table 12 grouped according to accident-type and period. The countermeasures evaluated are humps and corner refuges.

accident-type/ period	160*	311	312	410	510	610	660	871	872	876	878	ialt
before	1	1	2	1	7	1	4	1	1	1	0	20
after	0	0	3	1	5	3	3	0	0	0	1	16

* At the end of this paper a figure showing the full system of accident-types.

There is seen no change in the accident-numbers from before to after and no accident types (cfr. figure 8) differ from this result. As no change was seen from 1976 to 1977 in car-speeds there is good concordance between accident- and behavioural-trend.

The results of the accident-analysis can be found in part 4.3.16 of Notat 1/1983, Trafiksanering på Østerbro, part 1, Ulykkesanalyse.

2.3   Studies of pedestrians crossing Østerbrogade

As a part of the traffic restraint programme different kinds of centre-refuges was implemented in Østerbrogade.

To study the effect of these refuges behavioural studies of pedestrians were carried out on the northern and the southern part of Østerbrogade. The purpose of the study was also to gain knowledge of the location and frequency of crossing-behaviour in streets with pedestrian-tunnel, pedestrian-crossings ("zebras") and different refuges. Two working papers deal with this in detail, cfr. Engel (1982a & b).

The corresponding accident-analyses are found in section 4.3.10 and 4.3.14 in Notat 1/1983, Trafiksanering på Østerbro. Part 1 - Ulykkes-analyse.

2.3.1   Description of the accident-problem and the countermeasures
         implemented as prevention

In Østerbrogade several accidents with crossing pedestrians have oc-curred. Primarily in order to prevent these accidents a centre refuge was implemented on the northern part of Østerbrogade, cfr. figure 6 . The accidents were in the before-period of the types 811 and 812, cfr. figure 8 at the back. Another group of accidents were also tried prevented by this countermeasures. As these accidents did not involve pedestrains they are not treated here.

From Nøjsomhedsvej to Jagtvej the refuge is an "island" placed between two rows of kerb. The intention behind this is that primarily the pe-destrians get an area where they can wait while searching a possibili-ty for crossing the rest of the street as it is 22 metres wide at this section. By this arrangement the pedestrians can completely direct their attention towards one half of the street at a time. The pedestri-ans-crossing south of Koldinggade was in June 1973 substituted by a pedestrian-tunnel.

It should be mentioned that in 1973 there was implemented a centre refuge by means of painted lines on the street. The "real" refuge marked with kerbs were implemented in 1976-1977.

On the southern part from Olufsvej to Nøjsomhedsvej there has been collected accidents of types 811, 812 and 832 all involving pedestri-ans crossing the street, cfr. figure 8 . The refuge on the southern part is not as substantial as on the nothern part as it is split into several small "islands".

2.3.2   Description of the collection of behavioural data

Two studies are related to the behaviour of pedestrians. Both studies consist of counting as well pedestrians walking along the street as

pedestrians crossing the street. In addition the actual location and
course of the street-crossing of the pedestrians have been drawn. At
the same time it was recorded whether they stopped in the middle of
the street or not as whether they walked along the street in its middle.

Engel (1982b) has described a before/after-study related to the northern
part of Østerbrogade close to Koldinggade. The stretch of street
treated has a length of approximately 140 metres.

In June 1973 a pedestrians-tunnel connecting Koldinggade and Markens-
gade was opened.

October 2, 1974 there was made a before-registration of the pedestrians possibly using the tunnel and the centre-refuge at this time
consisting of painted lines. July 10, 1976 the after-recording of the
same behavioural data was done. At this time the centre refuge bordered
by kerbs was installed.

The observations were made from 7.30 to 11.18 and 14.30 to 18.18.
Each hour consisted of twice 15 minutes devoted to the registration
of crossing pedestrians and twice 6 minutes were used for the counting
of pedestrians walking along the pavement. In this way each day consisted of four hours registration of crossing pedestrians and 96 minutes counting pedestrians walking along the street.

Engel (1982a) describes the studies carried out on the southern
part of Østerbrogade close to Gustav Adolfs Gade. This is
purely an after-study as one was interested to know how the pedestrians
used the southern, more diffuse, centre refuge. An earlier study outside
the research area, at Sionsgade, guided the interest towards a study
of the proportion of pedestrians actually crossing the street and the
proportion using the pedestrians-crossing. The observations were done
June 19, 2977 using the time-scheme mentioned above.

2.3.3   Results concerning the crossing-behaviour of pedestrians

This analysis concerning the northern part of Østerbrogade has had its
pedestrian-counts scaled according to a time span of eight hours. The
figures are shown in table 15.

Table 15. Number of pedestrians walking along the pavement and crossing
the street (Østerbrogade at Koldinggade) grouped according to
period. The time of collection is 7.30 to 11.30 and 14.30 to
18.30, October 2nd 1974 and July 20th 1976*.

number of pedestrians/ period	pedestrians along the pavement	pedestrians crossing	proportion of crossing pedestrians in percentages
1974	4455	1506	33.83
1976	2875	826	18.73

* Inside this eight-hour-scheme crossing pedestrians were counted
during four hours while pedestrians walking along the pavement were
counted for 96 minutes. The figures in this table have been inflated
to the eight-hour-level.

It is seen that observations are given according to year and whether
the pedestrian crosses the street or not. There is a smaller propor-
tions of pedestrians crossing the street in 1976; this may be due to
the different months used in 1974 and 1976. The results of Sionsgade
show a crossing-proportion of app. 17%.

Table 16 show the proportion using the tunnel according to year of
observation. It is seen that the proportion of tunnel-users has decreased in 1976.

Table 16. Number of crossing pedestrians in Østerbrogade at Koldinggade
for 7.30 to 11.30 and 14.30 to 18.30, October 2nd 1974 and
July 7th 1976 grouped according to period and tunnel-useage*.

number of pedestrians/ period	crossing in tunnel	crossing in the street	proportion using tunnel
1974	265	488	35.24
1976	82	331	19.85

* The figures in this table are the actual counts collected in a four-
hour-period.

It is seen that the proportion of users in 1974 was 35%. The study
at Sionsagde showed a proportion of users of 82%. This might be due
to the fact that there is no centre refuge at this spot and the street
is narrower than the location at Koldinggade.

Finally it is seen from table 17 that in 1976 an increased proportion
uses the centre refuge to stop on or walk along. This proportion is
increased from 31 til 50%.

Table 17. Figures from table 16 (second column) grouped according to period and their useage of centre refuge.

period/ useage	striped centre refuge		refuge with kerbs	
	1974	%	1976	%
stopping	101	21	87	26
walking along refuge without stopping	51	10	79	24
crossing without stop	336	69	165	50
total	488	100	331	500

The study at the southern part of Østerbrogade at Gustav Adolfs Gade shows that during an hour the number of pedestrians walking along Østerbrogade is app. 328 and the number of crossing pedestrians is app. 69.

The proportion of crossing pedestrians is thus app. 21% while the corresponding figure at Sionsgade is 25%. It should be realized that the observed stretch at Sionsgade is 75 metres while it totals 110 at Gustav Adolfs Gade.

Table 18 gives the number of crossing pedestrians in each of the three parts of street. The observations are grouped according to the "manoeuvre", i.e. whether the pedestrians stops in the middle of the street and/or walk along the middel of the street.

Table 18. Number of crossing pedestrians in each of the three sections grouped according to usage of refuge and pedestrians-crossing. Figures are collected in Østerbrogade at Gustav Adolfs Gade, June 19th 1977.

street-section/ useage	pedestrians-crossing			
	1	2	3	total
stopping	2	99	9	110
walking along refuge without stop	2	2	1	5
crossing without stop	11	129	20	160
total	15	230	30	275

It is observed that 230 of the 275 crossing pedestrians (84%) use the pedestrians-crossing. This result is as well as the proportion crossing the street in good concordance with the results at Sionsgade.

Finally it is seen from table 18 that a greater proportion of pedestrians in the pedestrians-crossing make a stop in the middle while pedestrians not using the crossing do this in relatively fewer cases.

Comparing the two striped centre refuges at Koldinggade and Gustav Adolfsgade identical results are envisaged as the proportion of pedestrians crossing the street not in a right angle and without stopping is 31% at both locations.

2.3.4   Comparing the behavioural and the accident analysis results

In the follwing tables the accident figures are given for the northern and southern part of Østerbrogade separately as the refuges are different. It is seen from tables 19 and 20 that both types of refuges cause a decreased number of casualties and the decrease is of the same size independent of the type of refuge.

Table 19. Accidents involving pedestrians grouped according to period and type. The countermeasure is the kerbed refuge.

accident-type/ period	811*	812	total
before	6	7	13
after	1	1	2

Table 20. Accidents involving pedestrians grouped according to period and type. The countermeasure is the striped refuge.

accident-type/ period	811	812	832	total
before	5	2	1	8
after	1	0	0	1

* At the end of this paper is a figure showing the full system of accident-types.

As noted in the behavioural studies app. 20% of the pedestrians changed their crossing behaviour as intended and we thus find good concordance between accident and behavioural results.

It should be noted that the refuge on the southern part has seemed to introduce new accidents involving vehicles as these run into the small refuges introduced at this section.

The results of the accident-analyses can be found in <u>4.3.10 and 4.3.14</u>
<u>in Notat 1/1983, Trafiksanering på Østerbro. Part 1 - Ulykkesanalyse.</u>

Summary

The Danish Council of Road Safety Research has just finished its traffic replanning program in Østerbro, Copenhagen. This project has been
carried out as a before/after-study, each period consisting of three
years. A variety of countermeasures has been implemented and some have
proved effective and others have not. Related to some of these results
there has been studied behaviour of the street-users primarily to see
whether the behaviour changed as intended after the implementation of
the countermeasures.

The total number of studies as well as the three mentioned above show
good concordance between results of the accident analyses and the behavioural studies. By this the behavioural studies help one to know why
some countermeasures work while others do not.

3   Discussion

The basic idea behind behavioural studies of traffic is to know why
accidents do happen. Statistical accident analysis can only provide
numerical evidence and not very much about causes. In order to gain
knowledge about accident-causes and before/after-changes in street-
user-behaviour one has to carry out behavioural studies. In Denmark
at least we have some good material consisting of the accident-reports
made by the police. This is the verbal description of the accident
made by the policeman and based on the explanation of the accident involved persons and if available witnesses. The reading of these reports is another very important source giving possible accident causes.

Accident-causation is the key-word of the accident-researchers and
behavioural studies are a very important tool when finding and evaluating possible accident-causes.

4   Answers to items given in the inviting letter

a) Definitions relevant to your conflict technique:
This techniques is not a conflict-technique but studies the behaviour
of all relevant street-users.

b) Type of conflicts or road situations observed:
Road situations relevant, i.e. supposingly including the accident prone
behaviour related to the accident-problem studied.

c) Choice of observation means:
This is totally dependent of the accident-problems trying solved.

d) Description of data collection procedure and scoring form:
These methods are dependent on the observation means and consequently
dependent on the accident-problem trying solved.

e) Data treatment procedure:
The statistical methods consist for continuous variables of tests of
homogeneity of variances, t-tests, analysis of variance, regression-
analysis, analysis of residuals, test of normality-assumption etc.
Observations of discrete type are analysed by means of log-linear Pois-
son-models. Combining discrete and continuous variables is done by
using covariance-analysis.

f) Training procedure:
This consists of telling the observer(s) about the accident-problem,
the causation hypothesis and the technical equipment used . Consequently
almost no training is needed.

g) Choice of observation-periods:
Usually modules of approximately one working day are used. The before
and after-studies are done on exactly the same day of the year in
the follwing year(s).

h) Evaluation of your own technique: results, already obtained (re-
    liability, validity etc.):
The techniques have almost no reliability-problems as they are seldom
dependent on the observers. The results from the Østerbro-project
show good concordance between before/after-accident and before/after-
behavioural studies.

i) Anticipated modifications to take place during the Malmö experiment
and expected consequences:

The methods are totally flexible and choice of methods is dependent on
the accident-problems and their structure in the analysed Malmö-cros-
sings.

5   References

Engel, Ulla (1982a):
"Short term" and area-wide evaluation of safety measures implemented in
a residential areal named Østerbo. A case study.
Appearing in: OECD, Seminar on Short-term and Area-wide Evaluation of
Safety Measures.
Institute for Road Safety Research SWOV, Leidschendam, Holland.

Engel, Ulla (1982b):
Fodgængeres krydsning af Østerbrogade; deres anvendelse af fodgænger-
overgang samt afstribet midterhelle i 1977.
Danish Council of Road Safety Research, working paper no. 74, 4.9.3,
Gentofte, Denmark.

Engel, Ulla (1982c):
Fodgængeres krydsning af Østerbrogade; deres anvendelse af fodgænger-
tunnel samt midterhelle ved afstribning henholdsvis med kantsten, 1974
og 1976.
Danish Council of Road Safety Research, working paper no. 75, 4.9.3,
Gentofte, Denmark.

Engel, Ulla & Thomsen, Lars Krogsgård (1978a):
Effektmåling af bilisters og cyklisters adfærd før og efter etablering
af forskellige former for afmærkning i Ndr. Frihavnsgade, Århusgade og
Randersgade.
Danish Council of Road Safety Research, working paper no. 36, 4.9.3,

Engel, Ulla & Thomsen, Lars Krogsgård (1978b):
Effektmåling af bilernes hastighed før og efter etableringen af over-
kørsler ved syv sidegaders udmunding i overordnede gader.
Danish Council of Road Safety Research, working paper no. 37, 4.9.3,
Gentofte, Denmark.

Engel, Ulla & Thomsen, Lars Krogsgård (1979):
Dokumentation af dataindsamling. (Effektmåling af bilister og cyklis-
ters adfærd før og efter etablering af længde- og parkeringsafmærkning
i tre gader på Østerbro).
Danish Council of Road Safety Research, working paper no. 41, 4.9.3,
Gentofte, Denmark.

Engel, Ulla & Thomsen, Lars Krogsgård (1983a):
Trafiksanering på Østerbro. Sammenfatning. (Traffic-replanning in
Østerbro. Summary, with a summary in English).
Danish Council of Road Safety Research, Report 25, Gentofte, Denmark.

Engel, Ulla & Thomsen, Lars Krogsgård (1983b):
Trafiksanering på Østerbro. Part 1 - Ulykkesanalyse. (Traffic-replan-
ning i Østerbro. Part 1 - Accidentanalysis).
Danish Council of Road Safety Research, Memorandum 1/1983, Gentofte,
Denmark.

Engel, Ulla & Thomsen, Lars Krogsgård (1983c):
Trafiksanering på Østerbro. Part 2 - Adfærdsanalyse. (Traffic-replan-
ning in Østerbro. Part 2 - Analysis of behavioural studies).
Danish Council of Road Safety Research, Memorandum 2/1983, Gentofte,
Denmark.

Ingason, Thorir (1981):
Rekonstruktion af trafikulykker.
Instituttet for Veje, Trafik og Byplan, Danmarks tekniske Højskole,
Kgs. Lyngby, Denmark.

Thomsen, Lars Krogsgård (1979a):
Dokumentation af dataindsamling. (Måling af bilers hastighed før og ef-
ter etablering af overkørsler i otte sidegaders udmunding i overordnede
gader).
Danish Council of Road Safety Research, working paper no. 44, 4.9.3,
Gentofte, Denmark.

Thomsen, Lars Krogsgård (1979b):
Teoretisk bestemmelse af bilers nedbremsningsforløb og måling af dette
i tre sidegader.
Danish Council of Road Safety Research, working paper no. 45, 4.9.3,
Gentofte, Denmark.

Thomsen, Lars Krogsgård (1982a):
Short Term and Area-wide Evaluation of Safety Measures Implemented in
a Residential Area Named Østerbro. The Statistical Tools.
Appering in: OECD, Seminar on Short-term and Area-wide Evaluation of
Safety Measures.
Institut for Road Safety Research SWOV, Leidschendam, Holland.

Thomsen, Lars Krogsgård (1982b):
Afsluttende analyse af bilers hastigheder på fem lokaliteter gennem
tre år. (Snithastighedsmålinger).
Danish Council of Road Safety Research, working paper no. 66, 4.9.3,
Gentofte, Denmark.

Thomsen, Lars Krogsgård (1982c):
Afsluttende analyse af bilers hastigheder før og efter etablering af
overkørsler i otte sidegaders udmunding i overordnede gader. ("510").
Danish Council of Road Safety Research, working paper no. 80, 4.9.3,
Gentofte, Denmark.

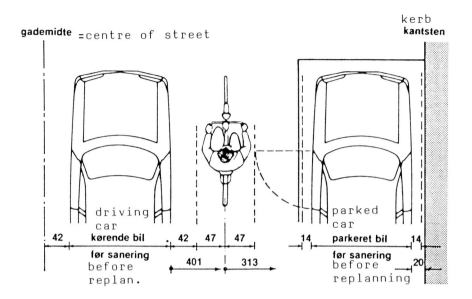

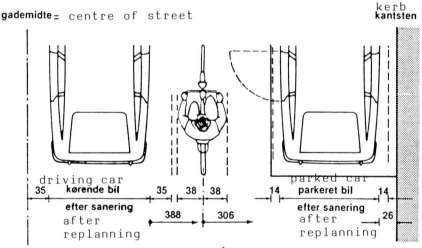

Figure 1. Distances in Århusgade before and the implementation of the restraint programme. This is an illustration of table 2.

ÅRHUSGADE

Rum for cyklister før sanering	221 cm =	space for bikes before r.
Rum for cyklister efter sanering	202 cm =	space for bikes after re.
Rum for cyklister difference	19 cm =	space-difference

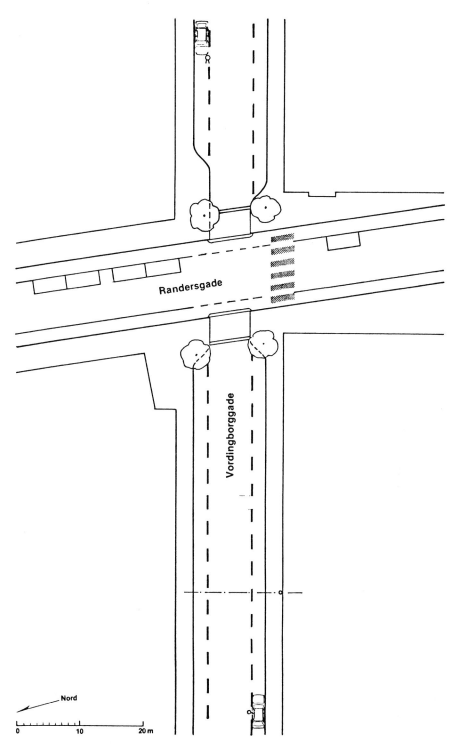

<u>Figure 2.</u> Colection of speed profiles at intersections
with humps and corner refuges.

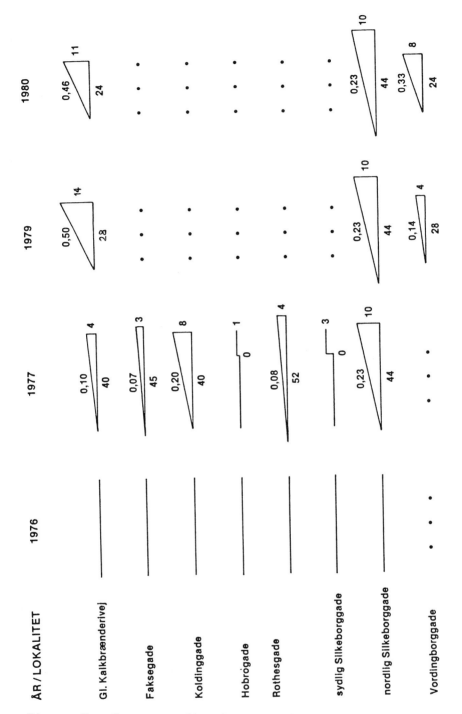

Figure 3. Corresponding humps and years of measurement.
Gl. Kalkbrænderivej 1979 and 1980 proved efficient.
All dimensions except the slope is given in centimetres.

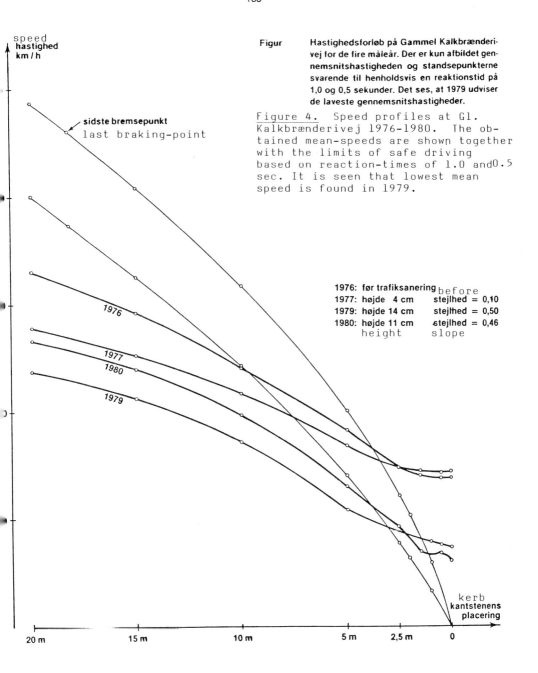

speed
hastighed
km / h

sidste bremsepunkt
last braking-point

Figur   Hastighedsforløb på Gammel Kalkbrænderi-
             vej for de fire måleår. Der er kun afbildet gen-
             nemsnitshastigheden og standsepunkterne
             svarende til henholdsvis en reaktionstid på
             1,0 og 0,5 sekunder. Det ses, at 1979 udviser
             de laveste gennemsnitshastigheder.

Figure 4. Speed profiles at Gl.
Kalkbrænderivej 1976-1980. The obtained mean-speeds are shown together
with the limits of safe driving
based on reaction-times of 1.0 and 0.5
sec. It is seen that lowest mean
speed is found in 1979.

1976: før trafiksanering before
1977: højde 4 cm   stejlhed = 0,10
1979: højde 14 cm   stejlhed = 0,50
1980: højde 11 cm   stejlhed = 0,46
        height   slope

1976
1977
1980
1979

kerb
kantstenens
placering

20 m   15 m   10 m   5 m   2,5 m   0

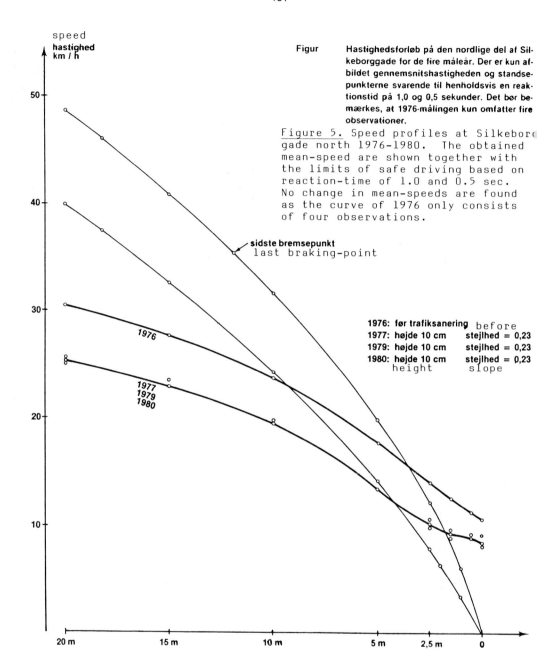

speed
hastighed
km / h

Figure 5. Speed profiles at Silkebor(
gade north 1976-1980. The obtained
mean-speed are shown together with
the limits of safe driving based on
reaction-time of 1.0 and 0.5 sec.
No change in mean-speeds are found
as the curve of 1976 only consists
of four observations.

sidste bremsepunkt
last braking-point

1976: før trafiksanering before
1977: højde 10 cm   stejlhed = 0,23
1979: højde 10 cm   stejlhed = 0,23
1980: højde 10 cm   stejlhed = 0,23
      height   slope

1976

1977
1979
1980

20 m   15 m   10 m   5 m   2,5 m   0

Figure 6.   Østerbrogade after the replanning with the kerbed centre refuge.

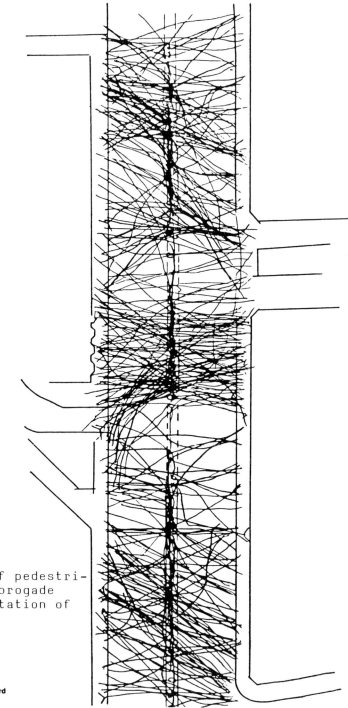

Figure 7. Study of pedestrians crossing Østerbrogade after the implementation of the centre refuge.

Nord

DANMARKS STATISTIK
Sejrøgade 11
Postboks 2550
2100 København Ø
Telf. (01) 29 82 22 - Telex 1 62 36

Figure 8. Accident-types used by the
Danish police and national bureau of
Færdselsuheld statistics.

Tælling
nr.

7200

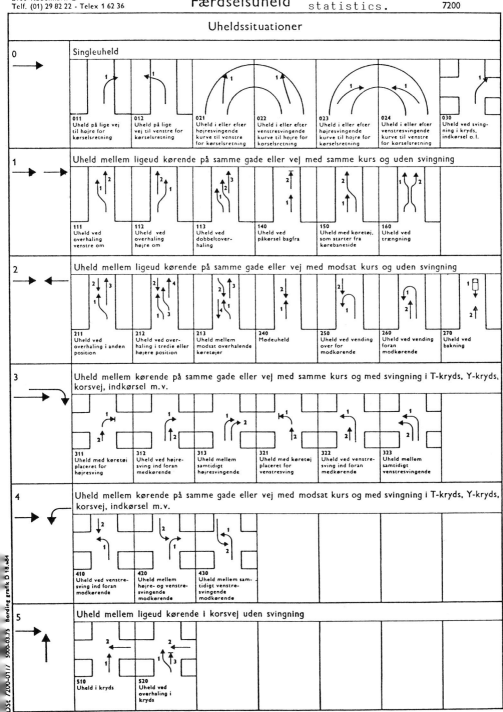

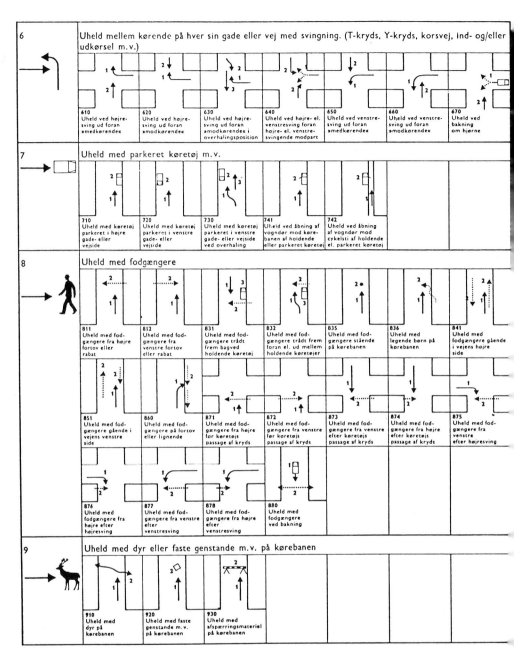

6 Uheld mellem kørende på hver sin gade eller vej med svingning. (T-kryds, Y-kryds, korsvej, ind- og/eller udkørsel m.v.)

| 610 Uheld ved højre-sving ud foran »medkørende« | 620 Uheld ved højre-sving ud foran »modkørende« | 630 Uheld ved højre-sving ud foran »modkørende« i overhalingsposition | 640 Uheld ved højre- el. venstresving foran højre- el. venstre-svingende modpart | 650 Uheld ved venstre-sving ud foran »medkørende« | 660 Uheld ved venstre-sving ud foran »modkørende« | 670 Uheld ved bakning om hjørne |

7 Uheld med parkeret køretøj m.v.

| 710 Uheld med køretøj parkeret i højre gade- eller vejside | 720 Uheld med køretøj parkeret i venstre gade- eller vejside | 730 Uheld med køretøj parkeret i venstre gade- eller vejside ved overhaling | 741 Uheld ved åbning af vogndør mod køre-banen af holdende eller parkeret køretøj | 742 Uheld ved åbning af vogndør mod cykelsti af holdende el. parkeret køretøj | | |

8 Uheld med fodgængere

811 Uheld med fod-gængere fra højre fortov eller rabat	812 Uheld med fod-gængere fra venstre fortov eller rabat	831 Uheld med fod-gængere trådt frem bagved holdende køretøj	832 Uheld med fod-gængere trådt frem foran el. ud mellem holdende køretøjer	835 Uheld med fod-gængere stående på kørebanen	836 Uheld med legende børn på kørebanen	841 Uheld med fodgængere gående i vejens højre side
851 Uheld med fod-gængere gående i vejens venstre side	860 Uheld med fod-gængere på fortov eller lignende	871 Uheld med fod-gængere fra højre før køretøjs passage af kryds	872 Uheld med fod-gængere fra venstre før køretøjs passage af kryds	873 Uheld med fod-gængere fra venstre efter køretøjs passage af kryds	874 Uheld med fod-gængere fra højre efter køretøjs passage af kryds	875 Uheld med fod-gængere fra venstre efter højresving
876 Uheld med fodgængere fra højre efter højresving	877 Uheld med fod-gængere fra venstre efter venstresving	878 Uheld med fod-gængere fra højre efter venstresving	880 Uheld med fodgængere ved bakning			

9 Uheld med dyr eller faste genstande m.v. på kørebanen

| 910 Uheld med dyr på kørebanen | 920 Uheld med faste genstande m.v. på kørebanen | 930 Uheld med afspærringsmateriel på kørebanen | | | | |

Figure 8 (continued).

REGISTRATION AND ANALYSIS OF TRAFFIC CONFLICTS BASED ON VIDEO

Richard van der Horst
Institute for Perception TNO
P.O. Box 23 3769 ZG
Soesterberg, The Netherlands

1. Introduction

During the last five years video-techniques for the unobtrusive observation and analysis of roaduser behaviour have been widely used. Especially for the evaluation of counter-measures or new road design elements the analysis of roaduser behaviour may be very helpful in understanding the functioning of the traffic process in relation with local characteristics. In this context our research is not restricted to the rare events like accidents and serious conflicts; also other behavioural aspects like speed, speed changes, path chosen, place of stopping, etc. are taken into consideration. After a short description of the method some applications with respect to interactions between roadusers will be discussed.

2. Definition and types of conflicts

To describe the danger involved in a traffic situation, Hayward (1972) defined the time-to-collision (TTC). This measure is the time for two vehicles to collide if they continue at their momentaneous speeds and on the same path. If the vehicles are not on a collision course the value of TTC is infinite. If they are on a collision course the TTC is finite and will decrease with time. An evasive action like decelerating and/or swerving may lead to a minimum value for TTC, which then increases to infinity again. The minimum TTC value now can be taken as a critical measure for the risk involved in an interaction between roadusers. Hayward suggests to use a minimum TTC value of 1.0 s as a good threshold. The definition of a conflict then is:
"A conflict is a traffic situation with a minimum TTC less than 1.0 s".
The threshold value of 1.0 s seems to be a rather arbitrary choice and could depend on the type of interaction (car-car or car-cyclist) or on different speedclasses. From the studies conducted so far, it appears that interactions with a minimum TTC-value greater than 1.5 s, in general do not substantially contribute to figures based on min TTC < 1.5 s. So 1.5 s is used as an upper limit in our studies, until now restricted to urban areas. In most of the manoeuvre-combinations the interaction between motorvehicle and cyclist or moped-rider is of main interest in our conflict studies. Three types can be distinguished:

NATO ASI Series, Vol. F5
International Calibration Study of Traffic Conflict Techniques
Edited by E. Asmussen
© Springer-Verlag Berlin Heidelberg 1984

a) car from/to minor road -- cyclist on priority street (mostly on a separate cycle-track),

b) cyclist from/to minor road -- car on main road and

c) right-turning car -- cyclist on main road.

3. Observation method

As mentioned in paragraph 2, the TTC measure is used for describing the interaction between roadusers. For the computation of TTC curves the measurement of motion and position parameters is necessary. For the objective quantification of several aspects of roaduser behaviour registration by means of _film_ or _video_ in most cases is still the only way. In a preliminary study both techniques were compared (Horst and Symonsma, 1979). With respect to costs and practical aspects the use of video is preferred.

For the recordings a suitable position for mounting the camera(s) has to be found in the neighbourhood of the location, preferably at a height of at least 4 m above the road surfaces, and as unobtrusively as possible. For all locations under investigation (until now about 40), a good camera position could be found rather easily in adjacent buildings or lampposts. Two types of video-recordings are made: one continuously on a timelapse videorecorder (VHS-system), mostly with a reduction factor of four (12.5 fields/s) or eight (6.25 fields/s), and the other on a normal speed video-recorder (U-matic system), 50 fields/s. The U-matic recorder is started by hand when a roaduser from a relevant direction arrives and stopped when the manoeuvre has taken place. At each location video recordings mostly are made during one day for six hours, 8-10 h, 12-14 h and 15.30-17.30 h. During these periods traffic counts are made for periods of five minutes, normally on the spot, sometimes afterwards from the timelapse recordings. For further details about the recording equipment see Horst (1982).

4. Data analysis

Those aspects, for which clear behavioural alternatives can be distinguished, are scored by individual observers directly from the video-recordings. Compared with a direct scoring by observers in the field, the possibility of repeating the scene, is seen as an advantage with regard to the reliability. A programmable electronic grid can be mixed into the video picture which, for example, enables a simple measurement of speed, path chosen, place of stopping, waiting time, passing time, etc. To describe the interaction between roadusers a complete quantitative analysis is done by means of special developed video-analysis equipment (for details see also Horst, 1982).

The quantitative analysis consists of selecting positions of some points of the vehicle on successive video stills by positioning electronic crosshairs. By means of transformation rules, based on at least four reference points, x and y-positions of the video-plane can be translated to positions on the plane of the street. By differentiating successive positions in time, the speed of the vehicle can be obtained. Four samples per second appeared to be a reasonable compromise between accuracy and duration of analysis.

For the computation of the TTC measure see Horst (1982). Two important steps in the computation are the decision whether the vehicles are on a collision course and if so, the calculation of the TTC value itself. In order to do this, accurate vehicle dimensions are required. Data of the dimensions of current types of cars are available.

In advance a preselection of manoeuvre-combinations for the quantitative analysis is made in order to reduce the total amount of work.

The number of conflicts (for example defined as the number of interactions with a minimum TTC of less than 1.5 s), related to an exposure measure E, gives a risk-index for two intersecting traffic streams. The exposure measure E is defined as:

$$E = \sqrt{I_i * I_j}$$

where I_i and I_j are the number of vehicles in stream i and j during a given period.

5. Applications

In two studies the seriousness of interactions between roadusers was measured by calculating minimum TTC values. The first one was an evaluation of the functioning of new road design elements (humps, hobbles, lane constrictions, special pavement, etc.) at fifteen non-signalised priority intersections on two demonstration cycleroutes in The Hague and Tilburg, two cities in The Netherlands. Several aspects of roaduser behaviour were studied in detail in order to compare:
a) the actual behaviour and the behaviour as intended by the designers for each experimental location,
b) the actual behaviour between experimental locations, and
c) the actual behaviour at the experimental locations and the behaviour at five control locations without special provisions.
See also Horst (1980, 1982).

The second study was finished recently (Horst, 1983). It concerns a before and after study of countermeasures related to cyclists at seven locations within the demonstration project on redesigning urban areas in Eindhoven and Rijswijk, also two cities in The Netherlands. Two types of countermeasures were investigated, first, measures for cyclists at intersections in the transition area between a redesigned area and main traffic roads and, second, the construction of special cycle lanes on the approach of signalised intersections at main traffic roads.

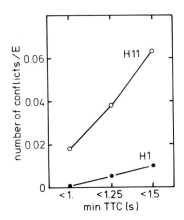

Fig. 1. Risk-indices (number of conflicts/E) based on different minimum TTC values for the experimental location (H1) and control location (H11).

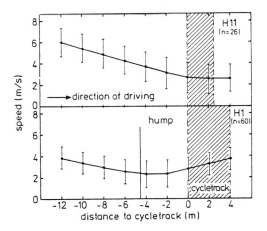

Fig. 2. Speed profiles (with standard-deviation) of crossing cars involved in interactions with cyclists on the cycletrack for an experimental location (H1) and control location (H11).

The method, based on objective quantification of interacting behaviour between roadusers enables not only a counting of traffic conflicts but also a process orient- ed analysis. For example Fig. 1 gives the risk-indices for the experimental location H1 (with a speed control hump at a distance of 4.5 m from the cycletrack) and the control location H11 (without special provisions), based on a minimum TTC of less than 1.5, 1.25 and 1 s, respectively.

An analysis of the speed profiles of crossing cars involved in the analysed interactions with cyclists at both intersections shows that this speed behaviour is largely responsible for the difference in risk-indices (Fig. 2).

At location H1 motorists pay more attention to the cycletrack, as indicated by the place were the minimum speed is reached, namely a few metres in front of the cycletrack instead of a few metres after the cycletrack for H11.

By comparing the behaviour between some experimental locations, the optimal place for the speed control hump in relation with the cycletrack could be determined, namely about 4 to 5 m in front of or after the cycletrack, instead of humps bordering the cycletrack.

In the demonstration project on redesigning urban areas one part of the study consisted of the analysis of cyclist's behaviour intersecting two high volume traffic streams. In the after situation a free zone with a width of two metres (with a re- fuge) has been constructed in the middle of the road, which could be helpful for cyclist crossing the main road. For several sections of the road, bordered by lines I to V (see Fig. 3) crossing times were measured from video and places of passing (in- dicated by sections A to F) were determined.

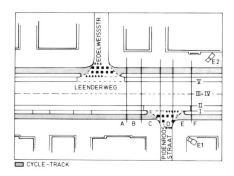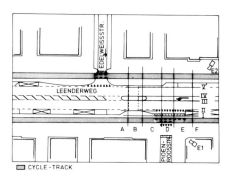

Before   After

Fig. 3. Situational map in before and after-situation for location E2 with grid for the analysis of the behaviour of crossing cyclists coming from the Pioenroos- straat.

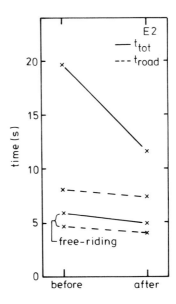

Fig. 4. Mean total crossing time ($t_{tot} = t_V - t_I$) and time on road ($t_{road} = t_V - t_{II}$) for cyclists coming from the Pioenroosstraat.

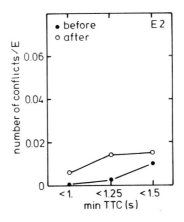

Fig. 5. Risk-indices based on different minimum TTC values in before and after-period at location E2 for cyclists coming from the Pioenroosstraat and cars on the main road.

Fig. 4 gives the results for all cyclists and also for the group of freeriding cyclists (without the presence of cars). t_{tot} is defined as the difference between the
passing time of line V and that of line I, while t_{road} is the difference between the
passing time of line V and that of line II. Before and after gives a difference of
t_{tot} of about 40%, caused by a decrease of waiting time in the zone between I and II,
just before the traffic lane is entered by the cyclists. Fig. 5 gives the risk-indices for the same manoeuvre-combination in before and after period.

In terms of conflicts in the after period cyclists take more risk then in the
before period, especially with respect to the second traffic stream. A further analysis might be a comparison of the gap-acceptance strategies of the cyclists in before
and after-period. This has not been done yet.

On the approach of a signalised intersection (location R1 Rijswijk) a special
cycle lane with a width of 1.5 m was constructed. For this measure the weave problem
between right-turn vehicles and cyclists on the cycle lane was thought to be of main
importance. For this type of interaction TTC curves were calculated. Fig. 6 gives the
results for before and after-period. Because of a recirculation measure the number of
right-turn vehicles at this intersection in the after-period is more than four times
as high as in the before-period.

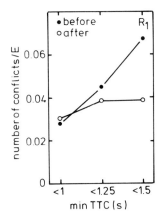

Fig. 6. Risk-indices for the weave type of interaction between right-turn vehicles
    and cyclist at location R1 without and with special cycle lane (width 1.5
    m).

That is the reason why the risk-indices, based on min TTC < 1 s, are the same, although the number of serious conflicts in the after period is twice as high.

It frequently results also in a long queue of vehicles waiting to turn right. This file is then blocking the cycle lane, which gives a lot of discomfort for the cyclists. For this type of situations, in which cyclists are manoeuvring between queues of waiting cars the calculation of time-to-collision curves is less meaningful because minimum TTC values are very low. Therefore, from the video recordings an additional classification into three groups was carried out, namely no, somewhat or serious hindrance for cyclists by waiting cars. The last category included stopping, getting off or swerving out over other traffic lanes or even over the footway by cyclists. In the after-period more than 20% of the cyclists met with this category, instead of only 2% in the before period.

Measurements of the free lateral space for cyclists near the stop line showed that the construction of the cycle lane (width 1.5 m) gives an extra mean free space of about 1 m.

6. Final remarks

The examples of paragraph 5 illustrate the power of a method, based on registration and analysis of roaduser behaviour by video. Such a method not only enables an objective registration of traffic conflicts, but gives also good possibilities for further analysis. However, at this moment it is not a practical operational tool, effective for research purposes but not for a widespread use. Therefore, at least a further automation of the analysis procedure is needed.

In principle for the Malmö study no particular adaptations of our method are expected. Usually in advance we restrict ourselves in the types of manoeuvre-combinations at the location which have to be recorded by video. Our observation periods are shorter than during the Malmö study. Also the number of traffic situations which can be analysed quantitatively is restricted for practical and financial reasons to about 200 in total.

7. REFERENCES

Hayward, J.Ch. (1972). Near miss determination through use of a scale of danger. Report no. TTSC 7115, The Pennsylvania State University, Pennsylvania.
Horst, A.R.A. van der and R.M.M. Symonsma (1979). Behavioural study by the Institute for Perception IZF-TNO. In: Older, S.J. & J. Shippey (Eds.). Proceedings of the 2nd International Traffic Conflicts Technique Workshop, May 1979. TRRL Suppl. Rep. 557, pp. 102-6. Transp. and Road Res. Lab., Crowthorne, Berkshire.
Horst, A.R.A. van der (1980). Behavioural study at the demonstration cycle routes at The Hague and Tilburg. Report IZF 1980 C-19, Institute for Perception TNO, Soesterberg (in Dutch).

Horst, A.R.A. van der (1982). The Analysis of Traffic Behaviour by video. In: Kraay, J.H. (Ed.). Proceedings of the third international workshop on traffic conflicts techniques, organised by the international committee on Traffic conflicts techniques ICTCT, Leidschendam, The Netherlands, April 1982. Report R-82-27, pp. 26-41. Institute for Road Safety Research SWOV, The Netherlands.

Horst, A.R.A. van der (1983). Demonstration project on redesigning urban areas: Behavioural observations in relation to bicycle traffic. Report IZF 1983 C-11, Institute for Perception TNO, Soesterberg (in Dutch).

JOINT INTERNATIONAL STUDY FOR THE CALIBRATION OF TRAFFIC CONFLICT TECHNIQUES

Background paper ICTCT Meeting Copenhagen, 25-27 May 1983 and Malmö,
30 May-10 June 1983

S. Oppe
Research psychologist
Institute for Road Safety Research SWOV
Leidschendam, The Netherlands

1. Introduction

In order to see why so much effort is made to calibrate traffic conflicts techniques, it is necessary to understand the fundamental ideas behind the analysis of conflicts.

Conflict observation is a way of looking at the unsafety of particular locations or situations in traffic.

Unsafety as such is not visible. We call a location unsafe if the probability
of an accident is too high. Accidents are rare events and seldom systematically observed. Accident potential is still harder to get at. We may arrive at a
statement about unsafety from several sources.

Sometimes a general theory about traffic safety is applied to a situation and
leads to statements such as "This particular lay-out of the intersection
causes too much risk to the cyclists coming from the right".

In this case the statement is assumed to be proven in general and applicable
to the situation under investigation.

In general, traffic safety theory is not that confirmed and statements like
the one above must be regarded as hypotheses that need confirmation. More
often one derives at the unsafety of a location from empirical evidence. The
frequency of accidents in the past is used to estimate the probability of an
accident.

In many cases, however, the accident frequency is too low to make reliable
estimates and additional information is then needed to get a more reliable
statement about the unsafety of a location.

Conflicts that are observed during some short period of time are often used as
if they were accidents in order to estimate the accident potential.

Therefore a conflicts technique is sometimes regarded as a surrogate measure
for accidents, used to detect the unsafety of locations or situations.

NATO ASI Series, Vol. F5
International Calibration Study of Traffic Conflict Techniques
Edited by E. Asmussen
© Springer-Verlag Berlin Heidelberg 1984

However, even if the conflicts techniques can be used for the detection of unsafety, then this is only the first step in the process of unsafety analyses. Much more important is what happens after the detection.

In order to improve safety, one has to analyse the problem and find the causes of unsafety and how these causes are provoked.

In most cases accident histories are scarce and far too incomplete to be used for these deductions, even if we use in-depth studies.

Observation of traffic behaviour at locations that are detected as dangerous (black) spots may clarify the safety problems and lead to effective safety measures.

The use of traffic conflicts techniques as behavioural observation techniques in safety analyses is completely different from the use of the techniques as a surrogate measure of the amount of traffic unsafety.

2. Definition of conflict behaviour

Traffic unsafety is the result of the various risks road-users meet, take or cause if they take part in traffic.

In general, traffic risk can be defined as the personal or material damage that may result from the decisions road-users take once meeting particular situations or other road-users.

In order to control risk, one has to know first which dangerous decisions take place and in which situation or under what circumstances these (conscious or inconscious) decisions are aroused or taken.

The study of dangerous traffic behaviour is fundamental for a good understanding of traffic unsafety. This study starts with the observation of behaviour and the context and the circumstances of that behaviour.

Because "the dangerousness" of the behaviour as such is not visible, an evaluation and interpretation of the situation is needed in order to detect "dangerous traffic behaviour".

Because of the subjectivity of such an evaluation and interpretation it is necessary to define narrow observation rules to arrive at objective data.

Scoring rules must be made explicit in such a way that there is an unambiguous mapping of cues in conflict severities. This entails more than agreement between observers only.

Traffic conflicts techniques may be regarded as techniques that enable systematic observation of dangerous traffic behaviour (conflicting behaviour). There are various conflicts techniques each using its own observation rules.

In order to compare results from different conflicts techniques, one has to know how the techniques have been used and from what kind of situations the data has arrived.

The first main question then is, what kind of situations a special investigator is concerned with, or stated otherwise, what is and is not a conflict. The second main question is, how dangerous was the situation he observed, or how serious is the conflict.

Defining a conflict, one may have different aims.

One may give a global demarcation of the concept and define the "universe of discourse". It becomes more interesting however if someone tries to give an operational definition of a conflict, in order to state the denotation of the concept instead of the connotation.

An operational definition is a rule to separate conflicts from non-conflicts. During the First International Workshop on Traffic Conflicts Techniques in Oslo, 1977, (Amundsen & Hydèn, 1977), conflicts were defined as:

"A traffic conflict is an observable situation in which two or more road users approach each other in space and time to such an extent that there is a risk of collision if their movements remain unchanged".

This definition seems to define the universe of discourse, but was primarily meant as an attempt to define a conflict operationally.

In fact, Perkins & Harris (1967) in their now classic paper, also used such a broad operational definition of a conflict. Their definition is unambiguous and easy to apply to car-to-car conflicts.

In practice, however, the conflicts techniques have been used with regard to various situations and each time a different operational definition has been given.

The following aspects are of importance:

- The investigation mostly regards only one aspect of traffic safety, e.g. the safety of children, pedestrians, intersections, serious accidents etc. Only those kinds of conflicting behaviour that are relevant for that aspect under consideration are classified.

- There is a variety of observation methods.

With more subjective methods we find terms such as "sudden behaviour" or "evasive action" as part of the definition, terms that presuppose a judgement of the observer. Objective methods use terms like "time-to-collision" (TTC) or "post-encroachment-time" (PET), terms that refer to registration apparatus.

- There is more or less differentiation in relevance of conflicting behaviour.

Terms like "serious" and "less-serious" conflicts have been used, referring to
the difference in accident potential. The seriousness dimension is seldom
specified - and if so - usually one dimensional (sudden action or not, short
or long TTC etc.).

Only in a few investigations we find more aspects, including qualitative aspects such as kind of road usage, to define the severity.

If we regard the conflict analysis technique as a systematic way of observation and investigation of risky interactive traffic behaviour, than the question what aspects of traffic behaviour are dangerous in which situations is
most important.

The usefulness of the conflict analysis technique does not, as it is often
stated, depend on the extent to which accident numbers are correctly predicted
but whether or not safety problems can be detected.

The prediction of accident numbers is often unrealistic due to the (statistic-
ly speaking) rare occurence. Validation of conflicts techniques with regard to
accident numbers will always be difficult, especially in situations where
there is no dense traffic. This kind of validation is not the exclusive one.
Another validation procedure that primarily regards the fundamental issues of
traffic unsafety is much more important. Attention must be stressed to the
confirmation of the conflict analysis technique as a theory about risky interacting traffic behaviour. Confirmation of a theory that tells us which
behaviour is dangerous in which situation.

To do this it is not enough to classify observations as conflicts. One has to
specify the seriousness of the conflict with regard to the accident that may
result from it. In order to do this, one has to state the relevant cues and
the weight these cues have with regard to the seriousness of the conflict.

The calibration experiment is planned as an international effort to arrive at
such a better understanding of danger in traffic. First we have to know what a
specific investigator is doing and how his doings are related to interactive
traffic behaviour. This is a premise to understand results from his work and
to relate these to one's own findings.

3. The seriousness of conflicts

If we take the seriousness of conflicts into account, then the problem of
finding a useful operational definition of a conflict, will be translated into
the assessment of the determinants of the conflict that are relevant with
regard to safety. The severity-rating is supposed to be a weighted sum of
these relevant determinants.

Knowledge of the relation between interactive traffic behaviour and safety is needed in order to state the degree of dangerousness of conflicts: the explanation of traffic unsafety in relation to traffic behaviour. Once this relation is stated, safety measures may be directed to the limitation or complete removal of serious conflict behaviour and the replacement by safe behaviour.

In depth studies of serious conflict behaviour as a tool in safety analysis are not yet well established.

In many cases one does not have the intention to accomplish a safety analysis, but as mentioned before one will use the conflict technique only to state the degree of unsafety of a location (absolute or relative, with regard to other locations).

However, also in this last case is the seriousness of conflicts of importance. We will give an example.

In Figure 1 the frequency distributions of conflicts are given for two locations. On the abcissa the degree of severity of the conflicts has been given. Let us define a specific kind of interactive behaviour to be a conflict if this behaviour is to the right of the point "conflict", and a serious conflict if it is to the right of the "serious-conflict" point etc.

We may notice that the following inclusions exist:

encounters $\supset$ conflicts $\supset$ serious conflicts $\supset$ accidents $\supset$ fatal accidents .

The area beneath the curve for location 1 at the right of the conflict point is equal to the total number of conflicts for location 1.

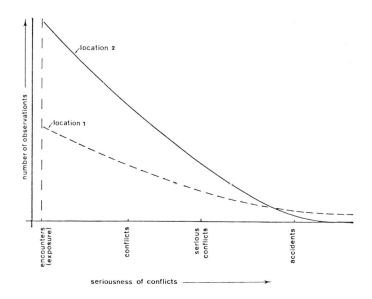

Figure 1.

If we estimate the relative safety of location 1 with regard to location 2
from the ratio between the numbers of conflicts of location 1 and location 2
then we will decide that location 2 is more dangerous.
If we use the serious conflicts both locations are almost equally dangerous.
If we use accidents, location 1 is more dangerous than location 2. Apart from
this it becomes clear from this figure that we try to estimate the area of the
very small right tail of the distribution from a very large portion of the
total area. Information about the shape of the curve is vital if we use these
estimates. This information is related to the validity and reliability of the
conflict technique.
It also is of importance for the relation between conflicts, exposure and
safety.

4. Conflicts, exposure and safety

We may wonder whether the picture of the two crossing curves in Figure 1 is
realistic or not.
If we restrict ourselves to accidents between road-users, then each encounter
between road-users may be regarded as a potential accident situation. The
total number of encounters may be used as a measure of exposure.
If the ratio between the number of encounters at both locations is equal to
the same ratio between conflicts, serious conflicts and accidents, then a
comparison between the safety of both locations will be done most reliably
using the (large number of) encounters. The relative unsafety is then directly
deduced from exposure.
If we compute for each location the accident rate (the ratio between the number of accidents and the measure of exposure, the number of encounters in this
case) and compare these rates for different locations then we are primarily
interested in the differences between the ratio of accidents for both locations and that ratio of the encounters. The picture of crossing curves corresponds to the idea of differentiating accident rates. A same kind of difference
may be expected with regard to the ratio's between the serious conflicts and
the conflicts.
This "conflict rate" will also give us information about the shapes of the
curves. If the conflict point and the serious conflict point are well-defined,
then we may use this rate, to state the relative unsafety of locations in a
more optimal way.

The more the conflict point equals the point of the encounters and the more
the serious conflict point reaches the accident point, the more the conflict
rate will resemble the accident rate. The difference in accident rate (or
conflict rate) will result in a less accurate prediction of accidents with
conflicts then with serious conflicts. If the conflict rate is equal for
various locations, then the validity of both measures should be equal and the
prediction of the number of accidents from the conflicts superior to the prediction based on serious conflicts because the former can be stated more
reliably. In general however, this will not be the case.
This dilemma between reliability and validity is important if we try to find a
useful conflict definition and we don't want to discriminate between conflicts
with regard to severity.
The choice of a definition will then be reduced to the problem of finding the
point at the severity scale that is optimal with regard to validity and
reliability.
In conclusion we may say that even if we want to predict the number of accidents or the degree of unsafety, we have to know what kind of interactive
behaviour is dangerous and how serious this danger is.

5. The calibration of conflict techniques

Each definition of conflicts and each scoring system of conflicts as used by
the different teams, is implicitely or explicitely based on a theory about
risky interactive behaviour. Some theories stress the subjective aspect of
this behaviour and try to evaluate the awareness of potential danger of the
participants in the conflict, some theories stress possibilities for correc-
ting behaviour in order to avoid an accident, some theories stress the possible consequences that may result if the conflict should become an accident.
These aspects are not independent of each other. Especially if the technique
is subjective and presupposes a judgement of the observer, then it is of importance to know what cues of the conflict situation are used and how the different cues are combined in order to get a final judgement. Whether a conflict
is serious or not, is not so much an empirical statement, but a theoretical
one.
All teams may learn from the confrontation of the theoretical points of view.
It is highly informative to know the similarities and dissimilarities of the
final judgements because these give us the operational discrepancy between
theories. Objective knowledge about the situations and especially the relation
between this information and the scoring system of each team may elucidate

discrepancies between scoring systems but also between theoretical views. This last kind of information is valuable not only for the application of conflict-techniques but also for a better understanding of traffic safety in general. The confrontation of traffic safety theories on its own, especially on an international scale, is reason enough to accomplish such an experiment.

With regard to the conflict techniques as such the main reason for doing the study is more specific.

As we know the justification for using the conflict technique depends on its reliability and validity. We have seen that these concepts depend on a proper definition of conflicts and are highly related to the proper severity scaling. In order to improve techniques and to compare results with those of other investigators or to interpret their results, one has to know what the other researcher exactly means by e.g. "serious conflicts".

Validation studies are very difficult to acomplish and are very expensive. In order to use the validation results of other investigators it is vital to know how to interprete their findings.

A comparison of the scales used for the determination of severity is also very important for this purpose. Calibration of conflicts techniques is the first step in the comparison of results and the exchange of ideas.

A comprehensive description of the analysis of the experimental data that must result in the information that is needed, has been given by Oppe (1982).

Literature

Amundsen, F.H. & Hydèn, C. (1977). Proceedings: First Workshop on Traffic Conflicts, Oslo 1977. Institute of Transport Economics, Oslo/Lund Institute of Technology, 1977.

Kraay, J.H. (ed) (1982). Proceedings of the Third International Workshop on Traffic Conflicts Techniques, Leidschendam, The Netherlands, 1982. R-82-27. SWOV, Leidschendam, 1982.

Perkins, S.R. & Harris, J.L. (1967). Traffic conflict characteristics; Accident potential at intersections. General Motors Corp., Warren, Mich., 1967.

Oppe, S. (1982). Statistical tools for the calibration of traffic conflicts techniques. R-82-37. SWOV, Leidschendam, 1982. Also in: Kraay (1982).

INTERNATIONAL CALIBRATION STUDY OF TRAFFIC CONFLICT TECHNIQUES

MALMO, 30 MAY - 10 JUNE 1983 : GENERAL DESIGN

ICTCT Steering Committee

Following discussions at the Copenhaguen preliminary meeting, agreement was reached on the detailed design of the calibration study and the data treatment procedure. These can be summarized as follows :

1. Data-collection

Ten conflict teams will take part in the experiment, nine of them using in average two field-observers, the last one working only from video-recordings.

Three locations have been picked up in the center of Malmö, to be observed simultaneously by all the teams ; two are non-signalized intersections and the third one is light-controlled. All three of them show a good traffic mixture (cars, two-wheelers and pedestrians) and are of a usual urban size, well adapted to observation by a normal conflict team.

Video-recordings will be made by two different video-systems, positionned in such a way as to give a good view of each junction. The observation-field for all observers on the ground will be the same as the camera-field.

The length of the data collection period adopted is 16 hours for each junction, between 7 a.m. and 8 p.m. In order to get sufficient conflict data for comparison purposes, part of the lunch-time and afternoon periods will be observed several times. Three days are necessary for data-collection on each junction and two hours is the maximum time that field-observers will have to work in a row.

A common "conflict data-sheet" has been agreed upon, to be used by all observers to record each conflict detected. Basic data for use in conflict labelling is pre-formated on the data-sheet, each team being free to add on to it according to their usual procedures. Data-sheets for all conflicts of the day are to be collected every evening, each team having checked beforehand that their data is correct (data-sheets properly filled in, one sheet and one only for each conflict).

The first half-day of data-collection will be a test-period as it will be necessary to check that the proposed procedure can actually work, and that possible practical problems can be solved.

NATO ASI Series, Vol. F5
International Calibration Study of Traffic Conflict Techniques
Edited by E. Asmussen
© Springer-Verlag Berlin Heidelberg 1984

2. Complementary data

Apart from conflict counts, some other data have been found necessary by many of the teams involved, so they can draw their own results.

It has been agreed that the Swedish and Danish team will handle accident information, and that the Danish technique of behavioural observation will provide some background data. Traffic counts will be performed from video-films.

3. Conflict labelling

Data treatment requires that all conflicts recorded by any team should be labelled, in order to get a proper reference sample. The labelling procedure will be performed by members of the participating teams not involved (if possible) in conflict observing. Every day, conflicts recorded during the day before will be labelled.

The "labelling team" will have in hands all the data-sheets and will check each event recorded on the video-films, using date and precise time, and description of manoeuvres and road-users involved. A list of conflicts labelled will be kept. Conflict numbers will be added to the original data-sheets which will be handed to S. Oppe, from SWOV, for statistical treatment. Photo-copies of their own data-sheets will be given back to each team after the labelling session.

Nos discussion will take place during the labelling sessions, every event recorded as conflict by at least one team being accepted as such.

4. Data treatment

A first statistical treatment will be performed in Malmö by S. Oppe and results will be presented on the last day of the study. Computer programmes to be used will have to be adapted on the computers available locally.

More in-depth statistical data analysis will be carried out at SWOV during the summer, and will provide comparisons between the participating TCTs with regards to conflict detection and severity scaling. Simultaneously, a detailed description of a sample of conflicts recorded by more than one team (the maximum sample size will be 100) will be performed at TNO on the basis of the Dutch video-technique. Results of both studies should be available at the end of November 1983.

Each participating team will also produce a report based on their own data ; content of these reports is left to the choice of the author, but elements of safety diagnoses on each junction, and critical observations about the experimental conditions will be welcome. These reports should be ready in October 1983.

5. <u>Results</u>

It will be necessary to get to an agreement on assessment of the calibration outcome and conclusions of the Malmö study. When results of detailed conflict comparisons and statistical analysis, as well as individual reports, are ready, a meeting of researchers (at least one per participating team) will be called. After this meeting an editing committee should have all necessary elements to produce a final comprehensive report and publish it. Editing is tentatively to be shared between Lund Institute of Technology, ONSER and SWOV. Final report should be ready in spring 1984.

COPENHAGEN ICTCT MEETING, May 25th - 27 th 1983 :

A SUMMARY OF DISCUSSIONS AND CONCLUSIONS

Nicole MUHLRAD
ICTCT Steering Committee

Discussions took place after each presentation and a wide range of both theoretical
and practical question were debated. These can be classified as follows :

1. The definitions relevant to the different TCTs

Most of the TCTs presented in Copenhaguen and calibrated in Malmö are based on the
definition of a conflict established in Oslo in 1977, that requires the existence
both of a collision course between two road-users and of an evasive action taken by
one of them. The Canadian technique, based on PET ("post-encroachment time") measurement, is an exception as the events recorded don't necessarily include any evasive
manoeuvre ; however, when an emergency action is observed, the corresponding event is
noted as a conflict even if the PET couldn't be measured.

Save for the Dutch video-technique, all TCTs now use, to a various extent, the
subjective judgement of observers.

All TCTs include a severity rating for each conflict recorded. However, what is
called a "light" conflict may vary a lot : for some conflict teams, a "light" conflict
is defined as the less severe form of conflict that may be related to accidents ; for
some others, a "light" conflict is an indicator of some traffic deficiency or some
form of road-user's behaviour, but is unlikely to be linked directly to injury-accidents ; such "light" conflicts appear then to be very similar to what is called in
certain TCTs "potential conflicts" or "encounters". The calibration study should help
to clarify these definitions and avoid further confusion between events that may not
be the same.

When "potential" conflicts or "encounters" are recorded, they seem to be considered
mostly as background data, but the particular use that is made of them is not defined
in any fixed way.

Some severity scales are relevant to the relationship between conflicts and injury-
producing accidents, some others take into account all collisions. In this latter
case, a risk matrix is generally built, linking the type of collision anticipated,
the type of road-users and the type of road-junction to the probability of occurrence
of an injury-producing accident. The number of variables and level of precision
introduced in this matrix must take into account the quality level (exhaustivity,

NATO ASI Series, Vol. F5
International Calibration Study of Traffic Conflict Techniques
Edited by E. Asmussen
© Springer-Verlag Berlin Heidelberg 1984

reliability) of the data-collection performed by the observers. Also, the weight put on very severe conflicts should not be exaggerated as such events are rare (though a lot less rare than injury-producing accidents) and the random factor can't be totally ignored.

It has been stressed that some of the less severe conflicts, when they involve particularly vulnerable road-users, may be more narrowly connected to injury-producing accidents than conflicts of a different type showing a higher degree of emergency. Some data, such as age of pedestrians or two-wheelers involved in conflicts, should therefore be more generally collected than they are now.

Some more questions have been raised when examining the various working-definitions of conflicts :
- some teams classify the road-users involved as "offended" or "offending" ; such definitions are often linked to local regulations (main road, secondary road, priority system, etc...) and vary from one team to another, which sometimes creates misunderstandings.

- including simple traffic-violations into the conflict data-collection has been considered in several countries, but experiments showed that these violations did not correlate with accidents. The general feeling is that, while traffic-violations can be recorded as background data, they do not correspond to the accepted definition of a conflict and shouldn't be added in the conflict-count.

- the "rear-end" type conflict appears to be a special case in many TCTs : either there is a specific problem for recording rear-end conflicts due to the working-definition or observation means chosen, or it is the expected relationship between those conflicts and injury-accidents that seems to vary according to the type of situation.

- there is a problem in dealing with "deliberate conflicts", i.e. situations where one road-user leaves volontarily a narrow margin between himself and another one or tries to force his way into traffic. Is the risk corresponding to such a situation as high as in an unvolontary conflict ? Should it be recorded as conflict or not ? Deliberate conflicts can introduce a bias in the calibration-study as the different teams of observers will have been trained according to the usual practice of road-users in their own countries...

2. The data-collection procedures

Most TCTs now specify that the conflict data-collection will be performed by observers on the ground. Video-recordings made during the observation periods are used, either to check or complement the conflict-data directly collected, or to provide tools for the training of new observers.

It is generally accepted that, in the present state of development of technical aids, TCTs can only be handed over to local authorities or road-administrations for operational use if the data recording is performed directly by observers on the ground. The Dutch video system that will be experimented in Malmö is more a tool than a TCT and is not yet operational. Further automation will be necessary before wider implementation becomes practical. The cost factor will then have to be considered.

In TCTs as they are applied now, much depends on the choice and the training of observers. It is clear that not everybody can be a reliable observer, depending on educational background and psychological characteristics, and the choice of the conflict team is very important for the scientific quality of research work. However, it is also clear that when a TCT in handed over for operational use, a compromise will often have to be found between the scientific quality of the results obtained (that requires a strict selection of observers) and the necessity of using the available man-power (even if the training of observers is not quite successful in eliminating all biases).

The usual practice for data-collection is what has been termed "sector-observation" : observers are staying in a fixed place while recording and watch a fixed area (generally part of a road-junction, sometimes specific traffic-flows or manoeuvres). One technique (the Netherlands) also includes "personal observation", with observers following specific road-users in an area and noting conflicts along the route ; such a practice has the advantage of giving an estimate of road-users exposure and has been particularly used in residential areas.

3. The validation of TCTs

Validation studies carried out on existing TCTs have so far produced encouraging, but partial results. Part of the difficulty of validating conflicts lies in the definition of a conceptual framework for safety indicators and of a proper methodology. It is clear that the development of TCTs has been made necessary because of the relative scarcity of injury-accident data as soon as local problems are tackled, and often also because of insufficient quality of the accident data-collection ; these two conditions are also, quite logically, what creates the most important problems in designing validation studies : how to relate a new indicator to and old one which is itself, in many situations, found inadequate ? It seems now that the validation process cannot be limited to finding simple correlations between conflicts and injury-accidents.

It has been noted that some sharp safety problems may introduce biases in a validation study if the corresponding locations are overrepresented in the study-sample : it is the case in situations where accidents result from a very specific combination of local factors, which appears too rarely to be taken into account in a normal (short) conflict observation period.

Many studies have been carried out to check relationships between conflicts, accidents, traffic-flows, subjective feeling of safety of road-users etc... There are strong indications that very "light" conflicts (or encounters or potential conflicts) are not generally related to injury-accidents, but maybe rather related to traffic flows and subjective feelings of safety. On the other hand, serious conflicts and injury-accidents have no direct relationship with traffic flows. So far, better correlations have been obtained between serious conflicts and injury-accidents.

One question arising when comparing accident-data and conflict data is whether all accidents should be considered or only those occurring during conflict recording hours. Total length of conflict observation periods on a junction for validation purposes is one of the important methodological choices.

4. The possible applications of traffic conflict techniques

Conflict data can be used in complement to accident data to help solve some specific safety problems or design new countermeasures.

Conflict data can also be used to replace missing or insufficient accident data for the detection of hazardous locations in urban areas, and, most important of all, for before-and-after evaluation studies when short-term results are needed.

Finally, conflicts can be used as a research tool to investigate some aspects of road-users' behaviour, and as an educational one.

When accident-data is insufficient to draw an adequate safety diagnosis on a junction, traffic-conflicts appear to be a better replacement tool than just using the qualified judgement of traffic engineers ; conflict data is also more accurate and objective than results obtained through public inquiries, as the residents or local road-users often tend to rate safety in an unreliable way. Evidence on these two points has been found in several countries (England, France, the Netherlands in particular).

When a TCT is used for evaluation purposes in replacement of injury-accident analysis, it must have been properly validated, i. e. showed to be an adequate measuring tool for traffic safety (some sort of relevance between conflicts and injury accidents must have been found). This is not a strict requirement, however, when conflicts are simply used as indicators of deficiencies in the traffic system.

For short-term evaluation studies, accident-data is practically never appropriate. Conflicts appear as the most promising replacement tool, as the data-collection technique has the advantage of being uniform and easier to handle than unformalized behavioural studies. Results obtained on the basis of a validated TCT will also carry more weight than behavioural observations, for which repeatability is hard to prove and which still raise many methodological problems.

Short-term assessment of countermeasures should be made in a descriptive way, as opposed to a purely quantitative one. A simple conflict count on before and after periods can only lead to a statement on a situation : it is not informative enough to help with the development of appropriate countermeasures or the improvement of already applied ones. A more analytical approach, with conflicts used as a diagnostic tool, is necessary in order to be able to compare different experiences and discover new possibilities for safety action.

Before-and-after evaluation should particularly be analytic and comprehensive in situations where a countermeasure will seriously affect traffic organisation and road-users' habits in an area. In such a case, a simple comparison of before-and-after conflict (or accident) numbers will never be sufficient to judge whether the countermeasures is successful or not ; moreover, background data obtained from traffic counts, speed measurements and behavioural observation may well prove a necessary complement to conflict (or accident) analysis to assess the new situation.

For research purposes, conflicts can be used to improve countermeasure design and get a better understanding of how these countermeasures work, and how they influence road-users' behaviour. Questions such as the local relationship between very light conflicts and the more serious ones related to accidents may be important to eliminate a particular accident situation.

As a research tool, TCTs can also be applied to driver's performance analysis, and one example of such an application (Austria) has showed the potentialities of the TCT in this field. Another opening for the future may be the use of conflict-data and conflict films or video-recordings for educational purposes ; research is only just starting on this point.

5. Conflicts as an operational tool

Conflicts are operational in a number of countries, where they are used on a wide scale by local authorities, the national road administration or the police (Great-Britain, Sweden, Germany, etc...) or recommended in guide-lines for traffic safety work. The general feeling is however that use of TCTs is still too limited and should be further developped.

In Sweden, the TCT is implemented by the National Road Administration on the assumption that there is a relationship between conflict and injury-accidents and that validation is going to be successful. Such a course of action has the double advantage of producing conflict data on a wide variety of locations, which will help validate the TCT, and also of ensuring that the TCT can be immediately used on a large scale as soon as its validity is recognized.

The general feeling is we are now confident enough in the value of TCTs for traffic safety work to promote operational use of it. We need large scale data, as well as assessment of countermeasures that have been designed at least partly on the basis of conflict data, and this cannot be obtained very easily if TCTs remain only a research tool.

6. The Malmö calibration study

It has been found important to calibrate all the available traffic conflicts techniques : accepted and calibrated TCTs open the way to international comparative studies of safety countermeasures ; also calibration will enable us to extend our national data-bases, by using data from other countries.

Calibration will also make it possible to improve our TCTs : we are still in a development phase, even though most techniques have been experimented over a number of years and further improvements should involve only minor changes. Some of the TCTs presented at the meeting are already operational, in their present state of development.

Calibration results and detailed data gathered from the Malmö experiment should be a precious help in designing adequate validation methods, through a better definition of the conceptual framework linking conflicts and accidents.

Finally, calibrated and accepted TCTs will be easier to promote in each of our countries for wider operational use.

Conclusion

Original TCTs developped in each country have been modified little by little, through direct experience as well as through comparisons with the work done in the same field by other teams of researchers. As a result, the different techniques seem to have been getting gradually closer to each other.

We have now reached a point where some TCTs are operational while others have reached a final stage of development, and there is a large amount of confidence in the validity of conflicts as a safety indicator. The Malmö calibration study will show whether the confidence is deserved and whether we all measure the same thing. From there on, validation work should become easier, and international cooperation in this field will have proved realistic, valuable and worth applying to this new issue.

Allowing for positive results of the Malmö International Calibration Study, three processes should be developping simultaneously in the near future :

- improving the existing techniques to increase validity and reliability (methods for training observers, recommendations for severity rating etc...)

- encouraging the operational use of conflicts in our various countries

- working on validation methodology and data-gathering for validation purposes.

Internationally accepted TCTs should be a starting point for an efficient cooperation in traffic safety research and in the design and evaluation of new traffic safety actions.

CLOSING REMARKS AT THE PREPARATORY MEETING FOR THE JOINT INTERNATIONAL CALIBRATION
STUDY OF TRAFFIC CONFLICT TECHNIQUES IN MALMO 25-27 MAY 1983

Prof. Erik Asmussen

director

Institute for Road Safety Research SWOV

Leidschendam

The Netherlands

Ladies and gentleman,

During the preparation of these closing remarks when Siem Oppe and Joop Kraay
discussed to give me an overall impression of the first two days of this meeting, I
suddenly got an association. The famous Einstein once said, and I quote not exactly:
what we as scientists or researchers see or observe from the real world is depending
on the characteristics of our measuring instruments. But these instruments are
designed in correspondence with the "a priory" theories we have. So, what we finally
observe from the real world is strongly depending on the knowledge we already have,
the theories we have at our disposal.

For outsiders this may give the impression that research is not more than a self-
fulfilling prophecy, biased by the way of thinking of the researcher.

Happily, this is not true because the most important characteristic of the researcher
is "his doubt" concerning the validity of his own findings and the findings of other
researchers.

In order to cope with this doubt, the researcher has developed methods to verify, to
falsify or to confirm findings, theories and methods.

One even can think that researchers are overcompensating their fear for the personal
bias.

Safety research is mostly an applied, interdisciplinary kind of research. Its function
is to give decision makers the information they need to select their strategies and
countermeasures to improve road safety.

Most of the decision makers in the western world are of opinion that the time is over
now for general or structural countermeasures, like safety belts, crash helmets and
also large scale infrastructural reconstructions. In their opinion further improvements
of road safety could be realised mainly by optimalisation of a great number of locations
or situations and by optimalisation of existing countermeasures, especially those
with the aim of influencing the road users' travel and traffic behaviour.

In order to do that they need an easy-to use technique to detect hazardous situations
and to carry out short term evaluation of countermeasures.

In some countries, f.i. in Holland they have gone so far that they use subjective

NATO ASI Series, Vol. F5
International Calibration Study of Traffic Conflict Techniques
Edited by E. Asmussen
© Springer-Verlag Berlin Heidelberg 1984

risk assessment of road users not only to detect hazardous situations, but also to evaluate the result of countermeasures.

However, as you all know there hardly is a correlation between verbal expressions and traffic behaviour, nor between the so called subjective (un)safety and the actual accident-rates of situations.

I think that the topic of this meeting, the development of conflict observation techniques, can give the decision makers a better alternative if they need more (and faster) information about traffic safety than accident figures.

And to be honest, how long ago we already gave the decision makers the impression that the very promising conflict technique could meet their needs for small-scale and short term desicion making and evaluation?

In the last five years they have only noticed little progression in the development of this technique, and above all they have noticed the lack of agreement between the researchers about which technique is the most appropriate to solve what problems.

The researchers working on this matter, were so fascinated by the new questions arroused by their own research that they became more or less isolated form their environment.

This envirionment, the decision makers, we must realise, have also the power to influence decisions about the money that is allocated for research.

Looking backward on this meeting and looking forward to the Malmö experiment in the context of the words I started with I consider this as a very good initiative both for researchers and decision makers.

I would like to congratulate all the members of the organizing committee and particulary Christen Hydén. Not only because he is the chariman of the ICTCT but also he was the "engine" behind both the meeting and the experiment.

I think that this meeting, as a preparation for the experiment, was a successful one. There were enthousiastic discussions, that thave lead to a better understanding of the differences fo the several techniques and especially of the thought behind them.

However, in practice the Malmö experiment will give the real advantages and disadvantages of the different techniques.

The aim of both the meeting and the experiment is in fact to convince the participants that they have developed not only reliable observation techniques relevant for the traffic safety problems, but also techniques applicable for operational use in the field for traffic engineeers and local authorities.

If we also want to convince the decision makers we need more than the proceedings of this meeting and the research results of the experiment.

Both should be integrated in a more comprehensive state of the art report and transmitted to all the decision makers concerned in this issue.

I am therefore very greatful that not only researchers attended this meeting.

On behalf of all the participants I whish to thank the members of the organising committee and I whish you all a very successful fieldstudy in Malmö.

ABSTRACTS

REVIEW OF TRAFFIC CONFLICT TECHNIQUE APPLICATIONS IN ISRAEL

A.S. HAKKERT
Road Safety Centre
Technion, Haîfa, Israël

This paper presents a brief description and overview of a number of TCT studies carried out in Israel by various researchers. First, a comparison between objective and subjective measures of traffic conflicts is described. Although some compatibility was achieved, overall agreement was not high. A second study described deals with a conflict study evaluation of flashing amber signal operation. Results of that study were encouraging and could warrant the conduct of a larger scale before-after experiment. Finally, a small operational application of TCT to a hazardous intersection is described. All studies were conducted at research organizations. The TCT in Israel cannot, as yet, be regarded in operational use.

CONFLICT OBSERVATION IN THEORY AND PRACTICE

V.A. GUTTINGER
"Advisie", The Hague, The Netherlands

The danger of traffic is commonly determined by the occurrence of accidents. This paper presents some of the history of alternative measures for describing traffic "unsafety": the measurement of so-called conflicts.

It not only goes into some theoretical problems concerning conflict definitions, it also summarizes the results of a series of research projects aimed at the development of a conflict observation technique for the estimation of the safety of pedestrians in residential areas.

The reliability, practical applicability and predictive validity of the developed technique, proved to be satisfactory.

It is concluded that the use of this technique seems to be justified for those situations in which accidents rates are relatively low; e.g. in residential areas. This is not only true because of the strong relationship between serious conflicts and accidents, but also because other potential alternative indicators for the estimation of traffic unsafety often used in practice (such as traffic volumes or subjective estimation of risk by residents), had little succes in predicting accidents.

At the end of the paper some of the results of the practical application of the developed technique are discussed.

Perhaps the face-validity and concurrent validity of the results of these applications of our observation technique, are even more convincing than the proven predictive validity.

NATO ASI Series, Vol. F5
International Calibration Study of Traffic Conflict Techniques
Edited by E. Asmussen
© Springer-Verlag Berlin Heidelberg 1984

TRAFFIC CONFLICTS IN BRITAIN : THE PAST AND THE FUTURE

G.B. GRAYSON
TRRL, Crowthorne, U.K.

This paper reviews the history of traffic conflicts research in Britain, and considers its possible development in the future. The results are discussed of work on the issues of reliability, repeatability, alternatives to conflict measures, and in particular the validity of conflict techniques. It is argued that the evidence is now such that the validity of conflicts can be regarded as established, in that conflicts can be shown to be related to accidents in an orderly and meaningful way.

The implications for future work are twofold. First there is the need for wider implementation at a local level. Second there is the prospect of using conflict studies as a research tool in the study of road user behaviour. In conclusion a plea is made for the better dissemination of the results of conflict studies in order to counter the criticisms that have been made of the technique.

THE DEVELOPMENT AND USAGE OF TRAFFIC CONFLICT TECHNIQUE ON THE SWEDISH ROAD NETWORK

M.O. MATTSON
National Road Administration, Sweden

The paper describes why the Swedish National Road Administration (SNRA) employs traffic conflict technique, the sort of traffic conflict technique used, and how it is applied.

At some traffic safety black-spots, the number of accidents are too few to allow application of the ordinary accident analysis in order to find the reasons for the accidents. In situations like these, traffic conflict technique is used as a supplement to the usual accident analysis.

The conflict technique used, was developed by the Lund Institute of Technology, primarily for urban areas. However, one element of this technique had to be altered, viz. the fixed time limit - 1.5 second - used to assess the gravity of the conflict, before it could be applied on rural roads. A substitution was made in the way that a speed-dependant factor was substituted for the fixed time limit.

The conflict technique can be applied all over the country since there is one official, trained in conflict technique, at each one of the 24 Regional Road Administations of the SNRA. The conflict technique is mainly used at locations with a rather limited area, e.g. intersections, where there are large enough traffic flows. The conflict data obtained is primarily used to supplement the accident data, in order to enhance the possibility of selecting the best measure for improving traffic safety.

THE TRAFFIC CONFLICT TECHNIQUE OF THE UNITED STATES OF AMERICA

J. MIGLETZ and W.D. GLAUZ
Midwest Research Institute,
Kansas City, USA

The traffic conflict technique has been practiced in the United States for over 15 years ; conflicts were first utilized on a large scale to solve operational problems at intersections. However, because there was a lack of a proven, direct relationship between accidents and conflicts, the US conflict technique has since received less emphasis from highway administrators.

New research has been carried out since 1979, with the aims of developping a standardized set of definitions and procedures for measuring traffic conflicts, and validating and calibrating the new technique. This paper describes the present American TCT : definitions, data collection, training of observers, evaluation, etc...

THE BRITISH TRAFFIC CONFLICT TECHNIQUE

C.J. BAGULEY
TRRL, Crowthorne, U.K.

A method of subjectively recording traffic conflicts or near accidents has been developed by TRRL for identifying safety problems and evaluating countermeasures, primarily at road junctions. This paper states the established conflict definitions and describes in detail the method of collecting conflict data and the way in which the data are used by many local highway authorities.

A training package for new observers is still under development and a brief synopsis of the contents of the training manual is included. The results of validation studies relating conflict counts to recorded injury accidents have been encouraging and a summary is given. Recently completed experiments in which conflict counts were compared with results from other subjective assessment methods at road junctions have demonstrated that these methods do not provide an acceptable alternative to conflict studies.

EXPERIENCE WITH TRAFFIC CONFLICTS IN CANADA WITH EMPHASIS ON "POST-ENCROACHMENT TIME" TECHNIQUES

P.J. COOPER
Insurance Corporation of British Columbia,
Vancouver, Canada

Following an initial investigation of the G.M. (brakelight) conflict recording technique in various Canadian cities in 1972/73, a study was undertaken to define conflicts in such a way that a better link between these and accidents could be established. The conflict definition arising from this work was referred to as

"Post-encroachment time" (PET). In the new technique, six conflict types are recorded by teams of three or four observers, each of them being attributed a specific task ; PET is timed for every event noted.

Evaluation of the PET technique was carried out from 1978 to 1981, but due to small data sample sizes (especially in terms of recorded accident frequency), results were inconclusive. One major positive finding, however, is that PET conflicts are generally better predictors of expected accidents than either past collision history or volume exposure factors.

THE FINNISH CONFLICT TECHNIQUE

R. KULMALA
Technical Research Centre of Finland
Espoo, Finland

According to the Finnish method, situations where braking or weaving begins 1.5 second or less before a potential collision are defined to be conflicts in urban traffic conditions, where the speed limit is 50 km/h. The time-to-collision value that defines conflicts varies with the speed limit or level at the study location. Also potential conflict situations are recorded in the Finnish method. Conflict observations are made by 2 - 4 persons. Video equipment is always used in order to check observations and to gather exposure and behaviour data. The method has been used in short-term evaluation of safety measures and safety analyses of junctions. The studies have been commissioned by Finnish towns, the Ministry of Transportation and the Roads and Waterways Administration.

THE TRAFFIC CONFLICT TECHNIQUE OF THE FEDERAL REPUBLIC OF GERMANY

H. ERKE
Technische Universität Braunschweig, Germany

The development of TCT started in 1973, was done by a group of traffic engineers and psychologists in Braunschweig (ERKE, GSTALTER, SCHWERDTFEGER, ZIMOLONG), and was based on general definition and observation procedures following PERKINS & HARRIS and SPICER. The observations that there are marked differences in behaviour (speed, acceleration, change in direction) and accidents (number, distribution, severity) between approaches and intersections led to the development of different types of observation. With these types of observation reliability and validity were sufficient. During the last years TCT was applied to the evaluation of safety measures by different research groups.

THE FRENCH CONFLICT TECHNIQUE

N. MUHLRAD, ONSER
G. DUPRE, CETE de Rouen
France

The French conflict technique has been developped at ONSER since 1973, primarily
with the aim of providing a tool for short-term evaluation of safety measures in
urban areas. Data-collection procedure is based on the subjective assessment of
traffic situations by field-observers (existence of a collision course, of an
evasive action, degree of emergency, etc...). A "risk-matrix", taking into account
type of manoeuvres performed, category of road-users involved and type of road
junction, links conflict data to the probability of injury-accidents occurring on
the studied location. A training manual has been issued ; observers' training is
both theoretical and practical and requires about three days, with necessary
checks on reliability during the first data collection period.

The French TCT has so far been mostly used in research, but some practical evaluation studies were also carried out, and potential use of conflicts for education
and information purposes is beeing considered.

THE SWEDISH TRAFFIC CONFLICT TECHNIQUE

C. HYDEN, L. LINDERHOLM
Lund Tekniska Högskdan, Sweden

Work with developing the Swedish technique started 1973 and a
technique for operational use was specified in 1974. Since then
modifications have been made and further developmental work is
still on-going.

Originally the following definition was adopted: A serious conflict
occurs when two road-users are involved in a conflict-situation
where a collision would have occured within 1.5 seconds if both
road-users involved had continued with unchanged speeds and directions. The recording of conflicts was, and still is, made by human
observers at the traffic site. Validation against police-reported
injury accidents was made. It was found that, out of many factors,
two had a definite influence on the relation between accidents
and serious conflicts. These are the kind of road-user involved
and the general speed-level at the intersection.

The original technique proved to work fairly well in operation but
has had some weaknesses. The most important ones are:

- The method did not give satisfying results for predicting
  accident risks in car-car situations, when dividing these
  into different types of accidents.
- The predicted risk of an accident for two identical conflicts
  could be different in different types of intersections.

In order to solve the problems mentioned two new definitions of a
serious conflict will be tested separately in an on-going project.

The first of these definitions is based on the hypothesis that, instead of a fixed threshold level at 1,5 seconds, this should be dependant on the actual speeds of the vehicles involved. The following definition is chosen:

A conflict is serious if the time-margin that remains when the evasive action is started is not more than the braking time at hard braking on slightly wet pavement plus half a second. The half of a second can be regarded as the remaining reaction margin.

The second definition is based on a subjective severity-scale where rate 1 corresponds to a conflict with a very small risk of a collision and rate 6 is a collision. Severity-rates 3 to 6 à priori define the serious conflicts. This subjective scale is introduced as a possible alternative because earlier work has shown that there is an observable difference between serious and non-serious conflicts.

The working procedure at the international calibration study in Malmö will be the same as normally used.

APPLICATION OF THE TRAFFIC CONFLICT TECHNIQUE IN AUSTRIA

R. RISSER, A. SCHUTZENHOFER
Austrian Road Safety Board

The paper contains a definition of traffic-conflicts which allows to differentiate between two types of conflicts : slight and serious ones.

The registration of conflicts in Austria, lately done while driving along in 200 subjects' cars in Vienna, aimed at judging drivers' performance, additionally to identifying dangerous spots in the road network.

The first training of observers, though, was carried out by means of video tapes and then on-the-spot. The third step, then, was to discuss events registered out of moving cars.

The Vienna study showed heterogenous results : Correlations between the accident record of the subjects and their conflict numbers on the standardized test-course were rather low (< 0,2) whereas the overall correlations between conflicts and accident numbers on the various sections of the test-course were fairly high (> 0,5).

Nevertheless, we found some types of erroneous behavior which in the past probably have led to higher accident numbers of some subjects. Those types of behavior were identified as errors in connection with too high a speed, too small distance to the preceding car, violations of traffic-light rules, stubborn behavior insisting on the right of way, risky overtaking.

The most important aim for the Malmö experiment was to find out,
if a person mostly used to conflict registrations out of moving
cars produces results comparable to those of other persons when
registrating traffic conflicts on-the-spot.

REGISTRATION AND ANALYSIS OF TRAFFIC CONFLICTS BASED ON VIDEO

R. VAN DER HORST
Institute for Perception TNO, The Netherlands

For the evaluation of counter-measures or new road design elements the analysis of
roaduser behaviour may be very helpful in understanding the functioning of the
traffic process in relation with local characteristics.
By means of unobtrusive observation of roaduser behaviour based on video an objective
quantification of behavioural aspects like speed, speed changes, path chosen, place
of stopping, etc. is carried out. For describing the danger involved in an interaction between roadusers the time-to-collision (TTC) is used.
After a short description of the method itself some applications are discussed, which
illustrate that the method not only enables an objective registration of traffic
conflicts, but also gives good possibilities for further analysis of the underlying
behaviour of roadusers.

THE USE OF TRAFFIC-BEHAVIOUR STUDIES IN DENMARK

U. ENGEL, L. THOMSEN
Danish Council of Road Safety Research

This paper deals with behavioural studies as a tool to evaluate safety measures. A
before-and-after study was carried out on a traffic replanning project in Osterbro,
Copenhaguen, each period consisting of three years. A variety of countermeasures
were implemented, some of them proved effective, others did not. In accordance
with some of these results, road-users' behaviour was studied, primarily to see
whether is had changed as intended after the implementation of the countermeasures.
The main items observed were drivers' behaviour related to street-markings, car
speeds, and pedestrians' behaviour in crossing some busy streets. The results
obtained were in good concordance with the accident analyses and it was showed
that behavioural studies were of help in understanding why some countermeasures
work while others do not.

JOINT INTERNATIONAL STUDY FOR THE CALIBRATION OF TRAFFIC CONFLICT TECHNIQUES :
BACKGROUND PAPER

S. OPPE
SWOV, Leidschendam, The Netherlands

When a statement on the unsafety of a location is needed, accident frequency is often too lows to make reliable estimates and additional data is then needed. Conflicts observed during some short period of time are often used as if they were past accidents to estimate the accident potential. The justification for so using a TCT depends on its relialibity and validity. These concepts are highly related to proper definition of conflicts and proper severity scaling. Validation studies are difficult to accomplish and very expensive. Calibration of conflict techniques is the first step in the process of validation and will be useful to design better studies and interpret future findings.

LIST OF PARTICIPANTS

---

1. Organising Committee

    Erik ASMUSSEN   Chairman
                                         SWOV, The Netherlands
    Ulla ENGEL   Organiser fo the meeting
                                           Danish Council of Road Safety Research, Denmark
    Christer HYDEN   Scientific Director
                                           Lund Tekniska Högskolan, Sweden
    Erdem IMRE   Administrator
                                           Swedish National Road Administration, Sweden
    Joop KRAAY   Treasurer
                                           SWOV, The Netherlands
    Nicole MUHLRAD   Technical Secretary
                                           ONSER, France
    Siem OPPE   Scientific Advisor
                                           SWOV, The Netherlands

2. Invited speakers

    Chris BAGULEY   TRRL, United Kingdom
    William T. BAKER   Federal Highway Administration, USA
    Peter COOPER   Insurance Corporation of British Columbia, Canada
    Guy DUPRE   Centre d'Etudes Techniques de l'Equipement, Rouen, France
    Heiner ERKE   Technische Universität Braunschweig, Germany
    Graham GRAYSON   TRRL, United Kingdom
    Viktor A. GUTTINGER   Advisie, The Netherlands
    S. HAKKERT   Technion, Haifa, Israël
    Risto KULMALA   VTT/TIE/AUR, Finland
    Leif LINDERHOLM   Lund Tekniska Högskolan, Sweden
    Ralf RISSER   Kuratorium für Verkehrssicherheit, Austria
    Richard VAN DER HORST   TNO, The Netherlands

3. Other speakers and participants

    Sverker ALMQVIST   Lund Tekniska Högskolan, Sweden
    Paul BAKKER   TNO, The Netherlands
    Lars EKMAN   Lund Tekniska Högskolan, Sweden
    Joop GOOS   Ministry of Transport, The Netherlands
    Dick IVARSSON   Trafiksäkerhetsverker, Sweden
    Mats-Ove MATTSON   Swedish National Road Administration, Sweden
    Jim MIGLETZ   Midwest Research Institute, USA
    Jean-François PEYTAVIN   ONSER, France
    Åse SVENSSON   Lund Tekniska Högskolan, Sweden
    Lars K. THOMSEN   Danish Council of Road Safety Research, Denmark
    Stefan ZABLOCKI   Malmö Gatukonror, Sweden

NATO ASI Series

Series F: Computer and Systems Sciences

Springer-Verlag
Berlin
Heidelberg
New York
Tokyo

No. 1

Issues in Acoustic Signal – Image Processing and Recognition

Editor: **C.H.Chen**
Published in cooperation with NATO Scientific Affairs Division
1983. VIII, 333 pages. ISBN 3-540-12192-7

Contents: Overview. – Pattern Recognition Processing. – Artificial Intelligence Approach. – Issues in Array Processing and Target Motion Analysis. – Underwater Channel Characterization. – Issues in Seismic Signal Processing. – Image Processing. – Report of Discussion Session on Unresolved Issues and Future Directions. – List of Participants.

This volume of the NATO ASI series is primarily concerned with underwater acoustic signal processing and seismic signal analysis, with a major effort made to link these topics with pattern recognition, image processing and artificial intelligence. The approach of artificial intelligence to acoustic signal analysis is completely new, as is the pattern recognition method to target motion analysis.

No. 2

Image Sequence Processing and Dynamic Scene Analysis

Editor:**T.S.Huang**
Published in cooperation with NATO Scientific Affairs Division
1983. IX, 749 pages. ISBN 3-540-11997-3

Contents: Overview. – Image Sequence Coding. – Scene Analysis and Industrial Applications. – Biomedical Applications.– Subject Index.

This volume contains the proceedings of a NATO Advanced Study Institute held 21 June – 2 July 1982 in Braunlage/Harz, Federal Republic of Germany, which was devoted to the rapidly emerging field of analyzing time-varying scenes and imagery. Twelve invited papers and twenty-six contributory papers cover a wide spectrum of topics which fall into three overlapping categories: displacement and motion estimation; pattern recognition and artificial intelligence techniques in dynamic scene analysis; and applications to diverse problems, including television bandwidth compression, target tracking, cloud pattern analysis, cell motion analysis and description, and analysis of heart wall motion for medical diagnosis. About half of the invited papers are tutorial overviews, while the rest – along with the contributory papers – describe the most recent progress in research. Together, they represent an invaluable reference tool for scientists and engineers working in time-varying imagery analysis and related areas, and perhaps the best single source of information for researchers just starting in the field.

NATO ASI Series

Series F: Computer and Systems Sciences

Springer-Verlag
Berlin
Heidelberg
New York
Tokyo

No. 3

Electronic Systems Effectiveness and Life Cycle Costing

Editor: **J.K.Skwirzynski**
Published in cooperation with NATO Scientific Affairs Division
1983. XVII, 732 pages. ISBN 3-540-12287-7

Contents: Mathematical Background and Techniques. – Reliability: Hardware. Software. – Life Cycle Costing and Spares Allocation. – Two Panel Discussions. – List of Lecturers and Delegates.

This volume contains the complete processings (including verbatim texts of several panel discussions) of a conference of the whole field of reliability and life cycle costing of modern electronic, computer-based systems. It contains a broad introduction to mathematical techniques supporting the field of technology, i.e. statistics, queuing theory, stochastic calculus, decision and utility theory. It presents reliability disciplines adopted by large organisations such as NASA, COMSAT, defence establishments in the USA and UK, in nuclear engineering and other areas. It concentrates both on hardware (including mechanical systems) and on software.
These proceedings are as an up-to-date statement of problems solved and problems encountered in predicting the behavior of electronic systems, their maintainability, spare allocation and operational costs. An especially important feature is the concentration on diagnosis of system malfunctions. Subjects covered in the panel discussions include: empirical prediction of failure rates; reliability and safety of computer based systems; hardware versus software reliability and maintainability; uncertainties in life cycle costing prediction; design audit programs; and aspects of warranties on system performance.

No. 4

Pictorial Data Analysis

Editor: **R.M.Haralick**
Published in cooperation with NATO Scientific Affairs Division
1983. VIII, 468 pages. ISBN 3-540-12288-5

Contents: Neighborhood Operators: An Outlook. – Linear Approximation of Quantized Thin Lines. – Quadtrees and Pyramids: Hierarchical Representation of Images. – Fast In-Place Processing of Pictorial Data. – C-Matrix, C-Filter: Applications to Human Chromosomes. – The Application of Gödel Numbers to Image Analysis and Pattern Recognition. – Segmentation of Digital Images Using a Priori Information About the Expected Image Contents. – A Syntactic-Semantic Approach to Pictorial Pattern Analysis. – Relational Matching. – Representation and Control in Vision. – Computer Vision Systems: Past, Present, and Future. – Artificial Intelligence: Making Computers More Useable. – Automation of Pap Smear Analysis: A Review and Status Report. – Medical Image Processing. – 2-D Fitting and Interpolation Applied to Image Distortion Analysis. – Pictorial Pattern Recognition for Industrial Inspection. – Pattern Recognition of Remotely Sensed Data. – Satellite Image Understanding Through Synthetic Images. – A Diffusion Model to Correct Multi-Spectral Images for the Path-Radiance Atmospheric Effect. – Analysis of Seasat-Synthetic Aperture Radar (SAR) Imagery of the Ocean Using Spatial Frequency Restoration Techniques (SFRT). – Adjacency Relationships in Aggregates of Crystal Profiles.